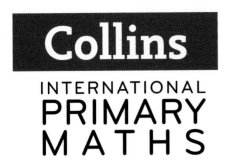

Collins

INTERNATIONAL PRIMARY MATHS

Student's Book 5

William Collins' dream of knowledge for all began with the publication of his first book in 1819. A self-educated mill worker, he not only enriched millions of lives, but also founded a flourishing publishing house. Today, staying true to this spirit, Collins books are packed with inspiration, innovation and practical expertise. They place you at the centre of a world of possibility and give you exactly what you need to explore it.

Collins. Freedom to teach.

Published by Collins
An imprint of HarperCollins*Publishers*
The News Building
1 London Bridge Street
London
SE1 9GF

HarperCollins*Publishers*
Macken House,
39/40 Mayor Street Upper,
Dublin 1,
D01 C9W8, Ireland

Browse the complete Collins catalogue at
www.collins.co.uk

10 9 8 7 6 5

ISBN 978-0-00-836943-9

British Library Cataloguing-in-Publication Data
A catalogue record for this publication is available from the British Library.

Author: Paul Hodge
Series editor: Peter Clarke
Publisher: Elaine Higgleton
Product developer: Holly Woolnough
Project manager: Mike Harman (Life Lines Editorial Services)
Development editor: Joan Miller
Copyeditor: Catherine Dakin
Proofreader: Tanya Solomons
Cover designer: Gordon MacGilp
Cover illustrator: Ann Paganuzzi
Typesetter: Ken Vail Graphic Design
Illustrators: Ann Paganuzzi and Ken Vail Graphic Design
Production controller: Lyndsey Rogers
Printed and bound in India by Replika Press Pvt. Ltd.

With thanks to the following teachers and schools for reviewing materials in development: Antara Banerjee, Calcutta International School; Hawar International School; Melissa Brobst, International School of Budapest; Rafaella Alexandrou, Pascal Primary Lefkosia; Maria Biglikoudi, Georgia Keravnou, Sotiria Leonidou and Niki Tzorzis, Pascal Primary School Lemessos; Taman Rama Intercultural School, Bali.

MIX
Paper | Supporting
responsible forestry
FSC™ C007454
FSC
www.fsc.org

This book is produced from independently certified FSC™ paper to ensure responsible forest management.

For more information visit: www.harpercollins.co.uk/green

The publishers gratefully acknowledge the permission granted to reproduce the copyright material in this book. Every effort has been made to trace copyright holders and to obtain their permission for the use of copyright material. The publishers will gladly receive any information enabling them to rectify any error or omission at the first opportunity.

Cambridge International copyright material in this publication is reproduced under licence and remains the intellectual property of Cambridge Assessment International Education.

Contents

Number

Geometry and Measure

Statistics and Probability

How to use this book

This book is used towards the start of a lesson when your teacher is explaining the mathematical ideas to the class.

Key words
• The **key words** to use during the lesson are given. It's important that you understand the meaning of each of these words.

• An **objective** explains what you should know, or be able to do, by the end of the lesson.

Let's learn

This section of the Student's Book page **teaches** you the main mathematical ideas of the lesson. It might include pictures or diagrams to help you **learn**.

 An activity that involves thinking and working mathematically.

An activity or question to discuss and complete in pairs.

Guided practice
Guided practice helps you to answer the questions in the Workbook. Your teacher will talk you through this question so that you can work independently with confidence on the Workbook pages.

HINT
Use the page in the Student's Book to help you answer the questions on the Workbook pages.

 Thinking and Working Mathematically (TWM) involves thinking about the mathematics you are doing to gain a deeper understanding of the idea, and to make connections with other ideas. The TWM star at the back of this book describes the 8 ways of working that make up TWM. It also gives you some sentence stems to help you to talk with others, challenge ideas and explain your reasoning.

At the back of the book

Lesson 1: Counting on and back (1)

- Count on and count back in steps of 7, 8 or 9

Key words
- count on
- count forward
- count back
- step size
- term

Let's learn

We can use a number line to help us count on or back in steps of the same size.

Count forward in 7s from 0.

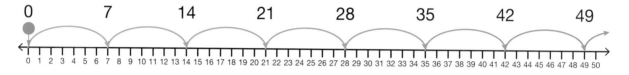

Count back in 9s from 185.

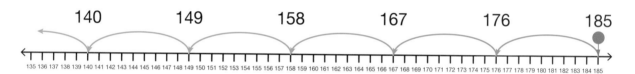

We can use an empty number line to count forward in 8s from 687.

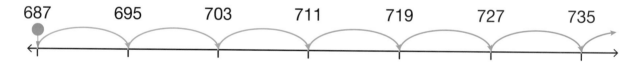

👥 Draw a number line to calculate.

$35 + 8 + 8 + 8 + 8$ $73 - 9 - 9 - 9$ $107 + 7 + 7 + 7 + 7 + 7$

Guided practice

Workmen construct a long underground pipe. They begin with a pipe 387 metres long. They build five new sections, each 9 metres in length. How long is the completed pipe?

387   396   405   414   423   432

432 metres

Lesson 2: **Counting on and back (2)**

Key words
- **count on**
- **count forward**
- **count back**
- **step size**
- **term**
- **negative number**

- Count on and count back in steps of a constant size from different numbers, including negative numbers

Let's learn

Count back in 5s from 14.

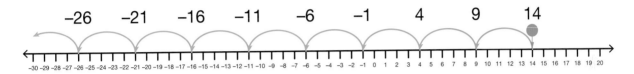

Count forward in 4s from –20.

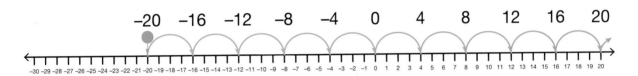

Count back in 6s from 17.

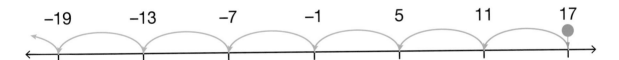

👥 A submarine goes down underwater at a rate of 8 metres per minute. If it starts at 0 metres below sea level, it will have a depth of –8 metres after 1 minute. What will its depth be after 2 minutes, 5 minutes and 9 minutes?

Guided practice

The temperature on Monday is 9 °C. The temperature falls by 4 °C every day. What will the temperature be on Sunday?

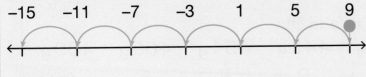

–15 °C

Lesson 3: **Number sequences**

- Find missing terms in a sequence

Key words
- sequence
- linear sequence
- term
- rule

Let's learn

A **linear sequence** is a number pattern that increases or decreases by the same amount each time.

The boxes of oranges below are arranged in a linear sequence. How many oranges are in the middle boxes?

10   ?   ?   ?   34

The difference between the start and end terms is 24 (34 – 10).

This is equal to four jumps of 6.

The middle boxes contain 16, 22 and 28 oranges.

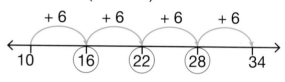

+ 6   + 6   + 6   + 6

10   16   22   28   34

Jin increases the time he runs by the same amount each day. On Monday he ran for 5 minutes and on Saturday he ran for 45 minutes.

How many minutes did Jin run for on Tuesday, Wednesday, Thursday and Friday?

Draw a diagram that explains the problem and how to solve it.

Guided practice

Seven buckets of sand labelled A to G are arranged in order of mass. The amount of sand increases by the same amount between buckets.

Bucket A has a mass of 415 g and bucket G has a mass of 469 g. What are the masses of the other buckets?

Difference between start and end terms: 469 – 415 = 54

Number of 'jumps': 6

Since 6 × 9 = 54, the rule is 'add 9'.

415 g	424 g	433 g	442 g	451 g	460 g	469 g
A	B	C	D	E	F	G

Number

Lesson 4: Square and triangular numbers

• Understand square and triangular numbers and extend sequences of these numbers

Let's learn

Square numbers
Numbers that form perfect squares

4   2 × 2 = 4

9   3 × 3 = 9

16   4 × 4 = 16

25   5 × 5 = 25

36   6 × 6 = 36

Triangular numbers
Numbers that form perfect triangles

1

3

6

10

15   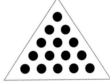

The multiplication square shows the first six triangular numbers.

Use the pattern to generalise and predict the next five triangular numbers.

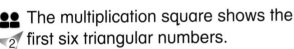

1	2	3	4	5	6	7	8	9	10	11	12
2	4	6	8	10	12	14	16	18	20	22	24
3	6	9	12	15	18	21	24	27	30	33	36
4	8	12	16	20	24	28	32	36	40	44	48
5	10	15	20	25	30	35	40	45	50	55	60
6	12	18	24	30	36	42	48	54	60	66	72
7	14	21	28	35	42	49	56	63	70	77	84
8	16	24	32	40	48	56	64	72	80	88	96
9	18	27	36	45	54	63	72	81	90	99	108
10	20	30	40	50	60	70	80	90	100	110	120
11	22	33	44	55	66	77	88	99	110	121	132
12	24	36	48	60	72	84	96	108	120	132	144

Guided practice

Grace arranges tennis balls to form triangular numbers. She needs 1 ball for the first term in the sequence. She needs 3 balls for the second term and 6 balls for the third term.

How many balls will she need for the seventh term?

1   3   6   10   15   21   (28)

Grace will need 28 tennis balls.

Number

Lesson 1: **Adding 2- and 3-digit numbers**

- Estimate and add 2- and 3-digit numbers, choosing the best mental or written method

Key words
- augend
- addend
- sum
- estimate
- expanded written method
- formal written method
- regrouping

Let's learn

$198 + 549 + 276 =$

Estimate by rounding: $200 + 500 + 300 = 1000$

Expanded written method

		1	9	8
		5	4	9
+		2	7	6
			2	3
		2	0	0
		8	0	0
	1	0	2	3

Formal written method

		1	9	8
		5	4	9
+		2	7	6
	1	0	2	3
		₂	₂	

Find the sum of each set of numbers. Estimate the answer first, and then use the expanded written method or the formal written method. However, check first to see whether a mental strategy is more appropriate.

     a 36, 45, 24   **b** 54, 66, 467

     c 367, 174, 248   **d** 437, 586, 734

Guided practice

Anwar takes four Maths tests and scores: 68, 77, 142 and 285. What is his total score?

Use the most appropriate strategy to find the sum. Estimate the answer first.

Estimate: | $70 + 80 + 140 + 290 = 580$ |

		6	8
		7	7
	1	4	2
+	2	8	5
		2	2
	2	5	0
	3	0	0
	5	7	2

Number

Lesson 2: **Adding 4-digit numbers**

- Estimate and add 4-digit numbers using the formal written method

Let's learn

3826 + 6746 =

Estimate by rounding: 3800 + 6700 = 10 500

Formal written method

- Put the digits into columns according to their place value.
- Draw answer lines under the last addend and write a + sign.
- Add the **ones** and regroup: 12 ones = 1 ten and 2 ones.
  Write 2 in the **ones** column and carry the 1 ten.
- Add the **tens** (including the carried 1 ten).
  Write 7 in the **tens** column.
- Add the hundreds and regroup:
  15 hundreds = 1 thousand and 5 hundreds.
  Write 5 in the **hundreds** column and carry the 1 thousand.
- Add the **thousands** (including the carried 1 thousand).
  Write 0 in the **thousands** column and 1 in the **ten thousands** column.

		3	8	2	6
+		6	7	4	6
	1	0	5	7	2
			1		1

Two runners complete a distance in three stages.

Runner A: 2346 metres, 5277 metres and 2858 metres

Runner B: 3455 metres, 4738 metres and 2676 metres

Does each runner complete a distance greater than 10 000 metres?

How do you know? Convince your partner.

Which runner completes the greater distance?

Guided practice

The balance of Mrs Yang's account is $7257. What is the new balance if she moves $8388 into her account?

Use the most appropriate strategy to find the total.
Estimate the answer first.

Estimate: $7300 + 8400 = 15 700$

		7	2	5	7
+		8	3	8	8
	1	5	6	4	5
			1	1	

Number

Lesson 3: **Adding positive and negative numbers**

• Use a number line to add a positive number to a negative number

Let's learn

Rule: Adding positive numbers to negative numbers – count forward the amount to be added.

Example: $-5 + 7 = 2$

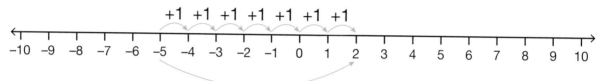

Example: $-8 + 3 = -5$

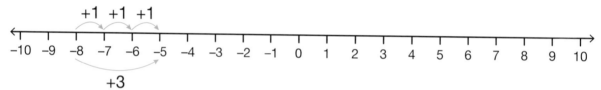

Work out the new temperatures after two rises in temperature.

Start temperature	−2 °C	−5 °C	−9 °C	−12 °C	−16 °C
Increase	1 degree	3 degrees	5 degrees	7 degrees	4 degrees
New temperature	_____°C	_____°C	_____°C	_____°C	_____°C
Increase	4 degrees	5 degrees	3 degrees	6 degrees	8 degrees
New temperature	_____°C	_____°C	_____°C	_____°C	_____°C

Guided practice

Mr Anand's account is overdrawn at −$9. This means he has spent more money than he actually had in his account.

He pays in $12. How much money does he have in his account now?

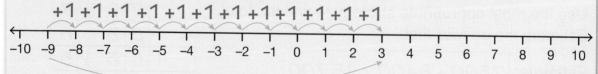

$-9 + 12 = 3$ Mr Anand now has $ 3 in his account.

Lesson 4: Identifying values for symbols in calculations

Key words
- unknown number
- symbol

- Find the value of unknown quantities represented by symbols in calculations

Let's learn

What numbers do the shapes represent?

$\square$ + 5 = 12 The square must represent 7 since 7 + 5 = 12 (or 12 − 5 = 7).

$\bigcirc$ − 4 = 10 The circle must represent 14 since 14 − 4 = 10 (or 10 + 4 = 14).

2 × $\triangle$ = 40 The triangle must represent 20 since 2 × 20 = 40 (or 40 ÷ 2 = 20).

Logan buys 3 pens for $9.

Grace buys 1 pen and 1 magazine for $7.

Write number sentences, using symbols, to represent the price of a pen and the price of a magazine.

What is the price of a pen? What is the price of a magazine?

Guided practice

2 identical red chairs have a combined mass of 20 kg.

1 red chair and 1 blue chair have a combined mass of 18 kg.

What is the mass of a red chair? What is the mass of a blue chair?

2 red chairs = 20 kg

1 red chair + 1 blue chair = 18 kg

$\square$ + $\square$ = 20 (since 2 × 10 = 20, $\square$ = 10)

$\square$ + $\bigcirc$ = 18

10 + $\bigcirc$ = 18 (since a red chair is 10 kg)

10 + 8 = 18 (since 10 + 8 = 18 or 18 − 10 = 8)

$\square$ = 10 and $\bigcirc$ = 8

The mass of a red chair is 10 kg and the mass of a blue chair is 8 kg.

Number

Lesson 1: **Subtracting 4-digit numbers (1)**

- Estimate and subtract 4-digit numbers, choosing the best mental or written method

Let's learn

$8362 - 4628 =$

Estimate by rounding: $8400 - 4600 = 3800$

Expanded written method

A

	8000	300	60	2
−	4000	600	20	8

B

			50	12
	8000	300	6̶0̶	2̶
−	4000	600	20	8
				4

C

			50	12
	8000	300	6̶0̶	2̶
−	4000	600	20	8
			30	4

D

	7000	1300	50	12
	8̶0̶0̶0̶	3̶0̶0̶	6̶0̶	2̶
−	4000	600	20	8
		700	30	4

E

	7000	1300	50	12
	8̶0̶0̶0̶	3̶0̶0̶	6̶0̶	2̶
−	4000	600	20	8
	3000	700	30	4

$3000 + 700 + 30 + 4 = 3734$

$8362 - 4628 = 3734$

Grace has answered four calculations but is worried that she has made some mistakes.

Check her calculations, using the expanded written method. Which answers are wrong? Discuss what you would say to Grace to help her avoid making these mistakes in future?

a $4684 - 1758 = 2926$

b $7358 - 3884 = 3574$

c $8445 - 2177 = 6268$

d $9567 - 4878 = 4589$

Guided practice

Janina and George play a video game. Janina scores 6348 points and George scores 3675 points. How many more points did Janina score than George?

Use the expanded written method to work out the answer. Estimate the answer first.

		5000	1200	140	
		6̶0̶0̶0̶	3̶0̶0̶	4̶0̶	8
−		3000	600	70	5
		2000	600	70	3

$2000 + 600 + 70 + 3 = 2673$

Estimate: $6300 - 3700 = 2600$

Janina scored 2673 more points than George.

Lesson 2: **Subtracting 4-digit numbers (2)**

Key words
- minuend
- subtrahend
- difference
- estimate
- formal written method
- regrouping

• Estimate and subtract 4-digit numbers using the formal written method

Let's learn

7268 – 3785 =

Estimate by rounding: 7300 – 3800 = 3500

Formal written method

- Put the digits into columns according to their place value.
- Draw answer lines under the subtrahend and write a – sign.
- Begin subtracting on the right (lowest place value) and work left, regrouping where necessary.
- For example, to allow subtraction to continue in the tens column (6 tens – 8 tens):
  - regroup 2 hundreds as 1 hundred + 1 hundred
  - move 1 hundred to the tens column to make 16 tens (1 hundred + 6 tens)
  - then 16 tens – 8 tens = 8 tens.
- Continue for the hundreds and thousands columns.
- Don't forget to cross through and rename digits that have been regrouped.

	$^6\not{7}$	$^{11}\not{2}$	^{16}	8
–	3	7	8	5
	3	4	8	3

Estimate which calculations have a difference:
  - less than 4873
  - greater than 4873

a 7444 – 2763 =

b 6256 – 1284 =

c 8533 – 3615 =

d 5946 – 1077 =

e 7287 – 2408 =

f 9314 – 4586 =

Guided practice

A large tank holds 5476 litres of water. A tap is opened and 2638 litres is poured out. What is the remaining volume of water in the tank?

Use the expanded written method to work out the answer. Estimate the answer first.

	$^4\not{5}$	^{14}	$^6\not{7}$	^{16}
–	2	6	3	8
	2	8	3	8

Estimate:

5500 – 2700 = 2800

The volume of water left in the tank is *2838* litres.

Number

Lesson 3: Subtraction where the answer is a negative number

- Subtracting pairs of numbers where the answer is a negative number

Let's learn

Rule: To subtract positive numbers from any integer, count back by the number to be subtracted.

Example: 5 − 7 = −2

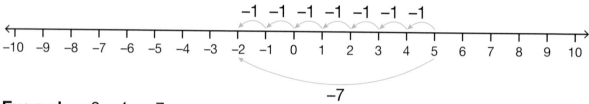

Example: −3 − 4 = −7

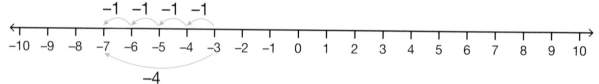

Work out the new temperatures after two falls in temperature.

Start temperature	3 °C	5 °C	8 °C	−4 °C	−6 °C
Decrease	4 degrees	7 degrees	10 degrees	1 degree	3 degrees
New temperature	_____ °C	_____ °C	_____ °C	_____ °C	_____ °C
Decrease	2 degrees	4 degrees	5 degrees	4 degrees	2 degrees
New temperature	_____ °C	_____ °C	_____ °C	_____ °C	_____ °C

Guided practice

Mr Onai has $4 in his account.

He uses his bank card to buy a book for $8. How much money does he have in your account now?

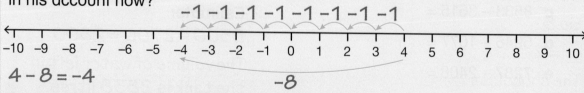

4 − 8 = −4

−8

Mr Onai now has −$4 in his account.

16

Lesson 4: **Identifying values for symbols in subtraction calculations**

- Find the value of unknown quantities in subtraction calculations that are represented by symbols

Let's learn

What numbers do the shapes represent?

$\square - 4 = 9$ The $\square$ must represent 13, since $13 - 4 = 9$.

Or you can use the inverse operation of addition: $\square = 9 + 4$
$= 13$

$15 - \bigcirc = 4$ The circle must represent 11, since $15 - 11 = 4$ (or $4 + 11 = 15$).

$\diamond + \diamond = 30$ The diamond must represent 15, since $2 \times 15 = 30$.

$\triangle - \diamond = 5$ Substitute the value of $\diamond$. This gives $\triangle - 15 = 5$.

The triangle must represent 20, since $20 - 15 = 5$ (or $5 + 15 = 20$).

Alfie visits a shop, where he spends $6 on a magazine.

He leaves the shop with $5.

How much money did he have before he bought the magazine?

Write a number sentence to represent Alfie's spending. Use a symbol to represent the amount Alfie had before buying the magazine.

Guided practice

Rajinda uses $10 to pay for a chocolate bar and receives $8 in change.

Lewis uses $10 to pay for a chocolate bar and a pencil set.
He receives $3 in change.

What is the price of the chocolate bar and the pencil set?

$10 - \square = 8$ (since $10 - 2 = 8$, $\square = 2$)

$10 - \square - \bigcirc = 3$

So $10 - 2 - \bigcirc = 3$, since I know a chocolate bar costs $2.

So $8 - \bigcirc = 3$. $\bigcirc$ must be 5 since $8 - 5 = 3$.

$\square = 2$ and $\bigcirc = 5$

The chocolate bar is $2 and the pencil set is $5.

Number

Lesson 1: **Prime and composite numbers (1)**

- Understand and explain the difference between prime and composite numbers

Let's learn

A number is **prime** if it has only two factors: itself and 1.

Numbers that have more than two factors are **composite** numbers.

The number 1 only has one factor, so it is not prime or composite.

Remember: A **factor** is a whole number that divides exactly into another whole number. For example, 1, 2, 3, 4, 6 and 12 are factors of 12.

Describe the number 22.

> Factors of 22: 1, 2, 11, 22
> 22 has more than two factors and so is **composite**.

Describe the number 19.

> Factors of 19: 1, 19
> 19 has only two factors, 1 and 19, and so is **prime**.

1 Find the missing digits to make each number prime.

   a 2 ☐   **b** ☐ 1   **c** 4 ☐   **d** ☐ 7

2 Find the missing digits to make each number composite.

   a 3 ☐   **b** ☐ 5   **c** 6 ☐   **d** ☐ 9

Guided practice

Is 48 a prime number? How do you know?

Since 48 has 8 factors other than 1 and itself, I know it is not prime.

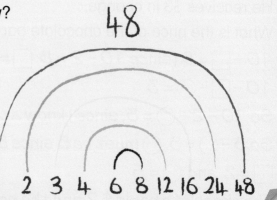

1 2 3 4 6 8 12 16 24 48

Number

Lesson 2: **Prime and composite numbers (2)**

• Understand and explain the difference between prime and composite numbers

Let's learn

The yellow squares in the table show **prime** numbers to 100.

	2	3	4	5	6	7	8	9	10
11	12	13	14	15	16	17	18	19	20
21	22	23	24	25	26	27	28	29	30
31	32	33	34	35	36	37	38	39	40
41	42	43	44	45	46	47	48	49	50
51	52	53	54	55	56	57	58	59	60
61	62	63	64	65	66	67	68	69	70
71	72	73	74	75	76	77	78	79	80
81	82	83	84	85	86	87	88	89	90
91	92	93	94	95	96	97	98	99	100

Eratosthenes

To prove a number less than 100 is prime, it is enough to show the number is not divisible by a 1-digit number other than 1.

Investigate whether this statement is true. Start by writing all the numbers up to 100 and marking the prime and composite numbers.

4

Guided practice

Draw a ring around the prime numbers. Use the diagram to help you.

9   (17)   27   (29)   49   77   (79)

It is not a
prime number.

Number

Does the number
divide exactly by
2, 3, 5 or 7?

Yes

No

It is a prime
number.

Number

Lesson 3: **Tests of divisibility: 4 and 8**

Key words
• factor
• divisible
• divisibility

• Identify numbers that are divisible by 4 and 8

Let's learn

A number is divisible by…	Example
2 if the last digit is an even number	4388 is divisible by 2, since 8 is divisible by 2
4 if the number formed by the last two digits is divisible by 4	7652 is divisible by 4, since 52 is divisible by 4
8 if the number formed by the last three digits is divisible by 8	9664 is divisible by 8, since 664 is divisible by 8

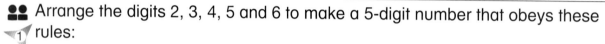

Arrange the digits 2, 3, 4, 5 and 6 to make a 5-digit number that obeys these rules:

• It is divisible by 5.

• When the final digit is removed, it becomes a 4-digit number that is divisible by 4.

• When the next final digit is removed, it becomes a 3-digit number that is divisible by 3.

• When the next final digit is removed, it becomes a 2-digit number that is divisible by 2.

Guided practice

Is the number 4676 divisible by 4? Is it divisible by 8?

I check divisibility by 4 by checking to see if the last two digits are divisible by 4.

$76 \div 4$
$= (40 \div 4) + (36 \div 4)$
$= 10 + 9 = 19$
Yes, 4676 is divisible by 4.

I check divisibility by 8 by checking to see if the last three digits are divisible by 8.

$676 \div 8$
$= (400 \div 8) + (276 \div 8)$
$= (400 \div 8) + (240 \div 8) + (36 \div 8)$
$= 50 + 30 + 4 \, r \, 4$
$= 84 \, r \, 4$
No, 4676 is not divisible by 8.

Lesson 4: **Square numbers**

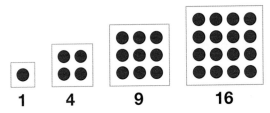

- Recognise square numbers from 1 to 100

Key words
- square number
- the square of

Let's learn

A **square** number is the product of a number multiplied by itself.

The square number sequence begins 1, 4, 9, 16…

1	4	9	16

The numbers are described as 'square' because they can be represented as square arrays.

Example: What is the square of 7? $7 \times 7 = 49$ or $7^2 = 49$

49 is a square number. It is the square of 7.

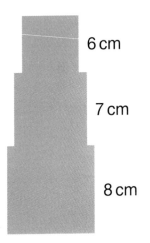

6 cm

7 cm

8 cm

A compound shape is made from three squares of sides 8 cm, 7 cm and 6 cm.

What is the total area of the shape?

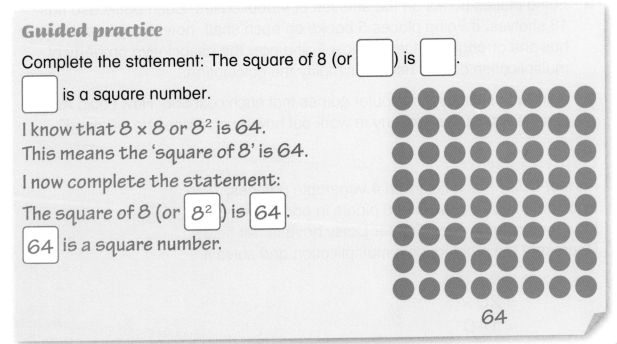

Guided practice

Complete the statement: The square of 8 (or ☐) is ☐.

☐ is a square number.

I know that 8×8 or 8^2 is 64.

This means the 'square of 8' is 64.

I now complete the statement:

The square of 8 (or 8^2) is 64.

64 is a square number.

64

Lesson 1: **Simplifying calculations (1)**

- Know which property of number to use to simplify calculations

Key words
- commutative property
- associative property
- distributive property

Let's learn

Multiplication has three properties:

1 a **commutative property**

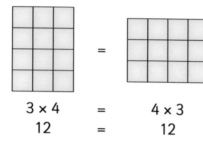

3 × 4	=	4 × 3
12	=	12

2 an **associative property**

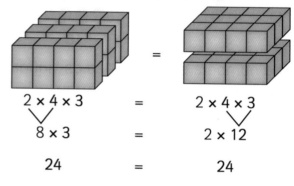

2 × 4 × 3	=	2 × 4 × 3
8 × 3	=	2 × 12
24	=	24

3 a **distributive property**

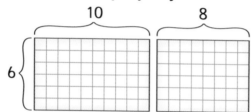

$6 × 18 = \boxed{6 × 10} + \boxed{6 × 8}$
$= 60 + 48$
$= 108$

a Faiha places books on the shelves of 4 bookcases. Each bookcase has 13 shelves. If Faiha places 5 books on each shelf, how many books has she arranged in total? Show Faiha how the associative property of multiplication can be used to simplify the calculation.

b Asim wants to buy 7 computer games that each cost $56. How could Asim use the distributive property to work out how much money he will need?

Guided practice

Farmer Daisy has a field with 4 vegetable patches.
Each patch has 7 rows with 5 plants in each row.
How many plants does Farmer Daisy have in her field?
Represent the problem as a multiplication and solve it.

$4 × 7 × 5 = 4 × 5 × 7$
$= 20 × 7$
$= 140$

Farmer Daisy has 140 plants.

Number

Lesson 2: **Simplifying calculations (2)**

• Know which property of number to use to simplify calculations

Let's learn

4 × 7

The **distributive** property means that a multiplication fact can be decomposed into the sum of two smaller facts.

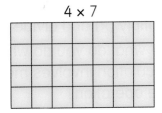

Break one of the factors apart

4 × 5   4 × 2

How would you use the distributive property of multiplication to solve these problems?

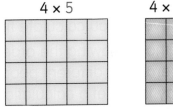

Multiply and add the products

a 67 × 6

c 88 × 8

b 74 × 7

d 96 × 6

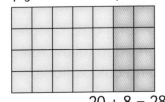

20 + 8 = 28

e There are 33 biscuits in each pack. If a family buys 6 packs, how many biscuits do they have altogether?

f Charlie sold 47 cups of lemonade each hour at her lemonade stand. She was at her stand for a total of 8 hours. How many cups of lemonade did she sell?

Guided practice

Use the distributive property to find the product.

$27 × 7 =$

$27 × 7 = (20 × 7) + (7 × 7)$

$\quad\quad = 140 + 49$

$\quad\quad = 189$

23

Lesson 3: **Order of operations (1)**

- Understand that the four operations of number follow a particular order

Key words
- numerical operation
- order of operations

Let's learn

Numerical operations include add (+), subtract (–), multiply (×) and divide (÷).

But what do you do when a calculation involves more than one operation?

$63 - 6 \times 8 =$

Which part of the calculation above do you calculate first?

Without knowing the order of operations, different people may interpret a calculation in different ways and come up with different answers.

So, long ago, people agreed to follow rules when doing calculations.

Order of operations

M × Multiplication
D ÷ Division
A + Addition
S – Subtraction

What is $63 - 6 \times 8 =$

First, multiply and/or divide. $63 - \boxed{6 \times 8}$

Then, add and/or subtract. $\boxed{63 - 48}$

So, $63 - 6 \times 8 = 15$ 15

👥 Perform the operations in the correct order and solve.

a $7 + 5 \times 2 =$ **b** $15 - 10 \div 5 =$ **c** $6 \times 3 - 2 =$

d $18 \div 3 + 6 =$ **e** $21 - 4 \times 2 =$ **f** $18 + 72 \div 9 =$

Guided practice

Use the order of operations to calculate.

a $21 + 35 \div 7 =$

$21 + \boxed{35 \div 7}$

$21 + 5$

26

b $12 \times 7 - 2 =$

$\boxed{12 \times 7} - 2$

$84 - 2$

82

Lesson 4: **Order of operations (2)**

- Understand that the four operations of number follow a particular order

Number

Let's learn

Some calculations involve more than just two numbers with one operation. Sometimes we need to carry out calculations where there are several different operations to do. It's very important that we know which order to calculate them so that we get the correct answer.

Order of operations

| M × Multiplication |
| D ÷ Division |
| A + Addition |
| S – Subtraction |

$20 - \mathbf{2 \times 5} = 20 - 10$
$\qquad = 10 \checkmark$

$8 + \mathbf{12 \div 4} = 8 + 3$
$\qquad = 11 \checkmark$

$\mathbf{18 \div 2} \times 3 = 9 \times 3$
$\qquad = 27 \checkmark$

$\mathbf{12 - 8} + 2 = 4 + 2$
$\qquad = 6 \checkmark$

$\mathbf{6 \times 2} + 4 = 12 + 4$
$\qquad = 16 \checkmark$

$20 - 2 \times 5 = 18 \times 5$
$\qquad = 90 \quad ✗$

$8 + 12 \div 4 = 20 \div 4$
$\qquad = 5 \quad ✗$

$18 \div 2 \times 3 = 18 \div 6$
$\qquad = 3 \quad ✗$

$12 - 8 + 2 = 12 - 10$
$\qquad = 2 \quad ✗$

$6 \times 2 + 4 = 6 \times 6$
$\qquad = 36 \quad ✗$

👥 Use the order of operations to calculate.

a $8 - 4 \div 2 =$

b $10 \times 9 + 11 =$

c $6 \div 2 + 4 =$

d $15 + 3 \times 5 =$

Guided practice

Use the order of operations to calculate: $50 - 15 \div 3$

By the order of operations,

I calculate the division first.

Then I calculate the subtraction.

The answer is 45.

$50 - 15 \div 3 = 50 - 5$
$\qquad = 45$

Number

Lesson 1: **Working with place value**

Key words
• place value
• product
• distributive property

• Use place value to multiply numbers to 1000 by 1-digit numbers

Let's learn

What is the best way to multiply?

786 × 6 =

786 × 6 is a little tricky to multiply in my head. I will use place value counters.

47 × 4 =

I can use the distributive rule to multiply 47 × 4 in my head.

47 × 4 = $\boxed{40 \times 4}$ + $\boxed{7 \times 4}$

= 160 + 28

= 188

×	700	80	6
6	100 100 100 100 100 100 100 100 100 100 100 100 100 100 100 100 100 100 100 100 100 100 100 100 100 100 100 100 100 100 100 100 100 100 100 100 100 100 100 100 100 100	10 10 10 10 10 10 10 10 10 10 10 10 10 10 10 10 10 10 10 10 10 10 10 10 10 10 10 10 10 10 10 10 10 10 10 10 10 10 10 10 10 10 10 10 10 10 10 10	1 1 1 1 1 1 1 1 1 1 1 1 1 1 1 1 1 1 1 1 1 1 1 1 1 1 1 1 1 1 1 1 1 1 1 1

4200 + 480 + 36 = 4716

👥 Rayah uses place value counters to multiply 263 by 8. How many 100s counters will she need? How many 10s counters? How many 1s counters? What is the product?

Guided practice

Draw the arrangement of place value counters to show the calculation and then use it to find the product.

8 × 453 = $\boxed{3624}$

100s	10s	1s
100 100 100 100 100 100 100 100 100 100 100 100 100 100 100 100 100 100 100 100 100 100 100 100 100 100 100 100 100 100 100 100	10 10 10 10 10 10 10 10 10 10 10 10 10 10 10 10 10 10 10 10 10 10 10 10 10 10 10 10 10 10 10 10 10 10 10 10 10 10 10 10	1 1 1 1 1 1 1 1 1 1 1 1 1 1 1 1 1 1 1 1 1 1 1 1
3200	400	24

Lesson 2: **Grid method**

- Use the grid method to multiply numbers to 1000 by 1-digit numbers

Key words
- **grid method**
- **partition**
- **product**

Number

Let's learn

Grid method

$376 \times 6 =$

Estimate by rounding: $400 \times 6 = 2400$

1 Draw a grid with three columns and one row (because you are multiplying a 3-digit number by a 1-digit number).

2 Partition the 3-digit number and write the hundreds, tens and ones across the top and the 1-digit down the side.

×	300	70	6
6			

3 Multiply each number by the 1-digit number and write the answers in the grid.

×	300	70	6
6	1800	420	36

4 Add the answers together to find the total.

$1800 + 420 + 36 = 2256$
So, $376 \times 6 = 2256$

 Look at these calculations.

A $4 \times 376 =$ **B** $3 \times 724 =$ **C** $6 \times 628 =$ **D** $8 \times 417 =$

Which of the calculations will give a product with:

    a the largest ones digit   **b** the smallest tens digit

    c the largest hundreds digit   **d** the smallest thousands digit?

Explain convincingly how you know.

Use the grid method to find the product of each calculation.

Guided practice

Estimate first, then use the grid method to work out the answer to the calculation. Show your working. Check your answer with your estimate.

$678 \times 6 =$ | 4068 |

Estimate: | $700 \times 6 = 4200$ |

×	600	70	8
6	3600	420	48

| $3600 + 420 + 48 = 4068$ |

Number

Lesson 3: Expanded written method

Key words
• expanded written method
• partition
• product

• Use the expanded written method to multiply numbers to 1000 by 1-digit numbers

Let's learn

Look at these three methods for multiplying a 3-digit number by a 1-digit number.

378 × 4 =

Estimate by rounding: 400 × 4 = 1600

Partitioning:

378 × 4 = $\boxed{300 \times 4}$ + $\boxed{70 \times 4}$ + $\boxed{8 \times 4}$

= 1200 + 280 + 32

= 1512

Grid method:

×	300	70	80	
4	1200	280	32	= 1512

Expanded written method:

```
        378
  ×       4
         32    8 × 4
        280    70 × 4
 +     1200    300 × 4
       1512
          1
```

👥 Critique the methods. Which of the methods do you prefer for working out the answer to 378 × 4?

Which is the most efficient method? Why?

Number

Lesson 4: **Real-life problems**

Key words
* grid method
* expanded written method

- Solve problems involving multiplication of numbers to 1000 by 1-digit numbers

Let's learn

Mrs Hernandez travels 7 hours by train every day.

How many hours will she travel in a year?

First, find the maths in the problem.

Daily travel time = 7 hours

Number of travel days = 1 year = 365 days

Total number of hours travelling = 365 × 7

Mrs Hernandez travels 2555 hours in a year.

Estimate by rounding:

$400 \times 7 = 2800$

		3	6	5	
×				7	
			3	5	5 × 7
		4	2	0	60 × 7
+	2	1	0	0	300 × 7
	2	5	5	5	

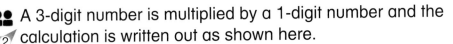

A 3-digit number is multiplied by a 1-digit number and the calculation is written out as shown here.

Each star stands for one digit.

The only digits that occur are 0, 1, 2, 3, and 7.

This is enough information to complete the whole multiplication.

Which digit does each star represent?

Clue: All the digits in the 3-digit number are the same!

		✻	✻	✻
×				✻
			✻	✻
		✻	✻	✻
+	✻	✻	✻	✻
	✻	✻	✻	✻

Guided practice

A large field is a regular octagon with sides 387 metres.

What is the perimeter of the field?

Number of sides = 8   Perimeter = 8 × 387 m

I estimate: 8 × 400 = 3200

×	300	80	7
8	2400	640	56

2400 + 640 + 56 = 3096

The perimeter of the field is 3096 metres.

Number

Lesson 1: **Working with place value**

• Use place value to multiply numbers to 1000 by 2-digit numbers

Let's learn

$36 \times 43 =$

Estimate by rounding: $40 \times 40 = 1600$

> As 30 is 10 times 3, the answer to 30×3 will be 10 times 3×3.

> As 30 is 10 times 3, the answer to 30×40 will be 10 times 3×40.

×	40	3
30	100 100 100 100	10 10 10
	100 100 100 100	10 10 10
	100 100 100 100	10 10 10
6	10 10 10 10	1 1 1
	10 10 10 10	1 1 1
	10 10 10 10	1 1 1
	10 10 10 10	1 1 1
	10 10 10 10	1 1 1
	10 10 10 10	1 1 1

$36 \times 43 = \boxed{30 \times 40} + \boxed{30 \times 3} + \boxed{6 \times 40} + \boxed{6 \times 3}$

$\qquad = 1200 + 90 + 240 + 18$

$\qquad = 1548$

So, $36 \times 43 = 1548$

👥 Olly wants to multiply 374 by 80. He is worried that he will not have enough place value counters for 80 rows of multiplication.

Explain to Olly how he can multiply without using 80 rows of counters

What is the product?

Guided practice

Use the area diagram to calculate the product.

32×28 $\boxed{896}$

×	20	8
30	$30 \times 20 = 600$	$30 \times 8 = 240$
2	$2 \times 20 = 40$	$2 \times 8 = 16$

$\boxed{600 + 240 + 40 + 16 = 896}$

Lesson 2: **Working with place value (2)**

Key words
- **dividend**
- **divisor**
- **quotient**
- **place value**
- **chunking method**

- Use place value to divide numbers to 100 by 1-digit numbers

Number

Let's learn

Using place value counters to represent a 2-digit dividend, you can model division by a 'chunking' method.

$95 \div 5 =$　　　　Estimate by rounding: $100 \div 5 = 20$

Step 1

10s	1s
10 10 10	1 1 1 1 1
10 10 10	
10 10 10	

Step 2

10s	1s
10 10 10	1 1 1 1 1
10 10 10	
10 10 10	Regroup four 10s as forty 1s.

1 group of 50 (or **10** groups of 5)

Step 3

10s	1s
10 10 10	1 1 1 1 1 1 1 1
10 10	1 1 1 1 1 1 1 1
	1 1 1 1 1 1 1 1
	1 1 1 1 1 1 1 1
	1 1 1 1 1 1 1 1
	1 1 1 1 1

Step 4

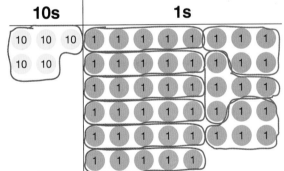

9 groups of 5

$95 = $ **10** groups of $5 + $ **9** groups of 5

$\quad\quad = 19$ groups of 5

$95 \div 5 = 19$

Use your 100s, 10s, 1s place value chart and place value counters to work out $84 \div 3$. Discuss each step with your partner.

Guided practice

Fill in the missing numbers and find the quotient.

$52 \div 4 = \boxed{40 \div 4} + \boxed{12 \div 4} = \boxed{10 + 3} = \boxed{13}$

35

Lesson 3: **Expanded written method**

Key words
- dividend
- divisor
- quotient
- expanded written method
- chunking method

- Use the expanded written method to divide numbers to 100 by 1-digit numbers

Let's learn

The expanded written method for division involves repeated subtraction of multiples or 'chunks' of the divisor.

Step 1

```
      2
  4 )9 1
```

Step 2

```
      2
  4 )9 1
      8 0    4 × 20
```

Step 3

```
      2
  4 )9 1
  -  8 0    4 × 20
     1 1
```

Step 4

```
     2 2
  4 )9 1
  -  8 0    4 × 20
     1 1
```

Step 5

```
     2 2
  4 )9 1
  -  8 0    4 × 20
     1 1
       8    4 × 2
```

Step 6

```
     2 2 r 3
  4 )9 1
  -  8 0    4 × 20
     1 1
  -     8    4 × 2
        3
```

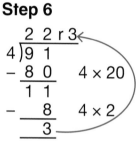 Riva says: 'There is only one 2-digit number that gives a remainder of 1 when divided by any of the numbers 5, 4, 3 and 2.'

Is she correct? If so, what is the number?

Describe the method you used to work this out.

Guided practice

Estimate first, then use the expanded written method of division to work out the answer to the calculation.

$98 ÷ 5 = \boxed{19\ r\ 3}$

Estimate: $\boxed{100 ÷ 5 = 20}$

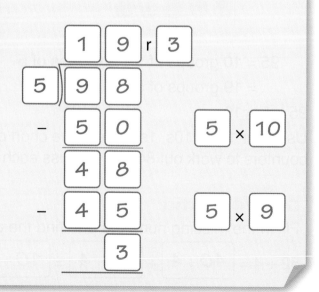

```
        1   9  r 3
   5 )  9   8
   -    5   0        5 × 10
        4   8
   -    4   5        5 × 9
            3
```

Lesson 4: **Real-life problems**

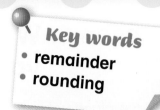

Key words
• **remainder**
• **rounding**

• Solve problems involving division of numbers to 100 by 1-digit numbers

Let's learn

For a division calculation, you usually just state the remainder (if there is one).

However, for some word problems, you may need to **round** the answer.

A carton contains 6 eggs.

a How many cartons are needed to pack 75 eggs?
$75 \div 6 = 12 \text{ r } 3$
The remaining 3 eggs still need to be put into boxes.
So, **round up**: 12 r 3 becomes 13, so 13 boxes are needed.

b How many boxes will be full?
3 eggs do not make a full box.
Ignore the remainder and **round down**: 12 boxes will be full.

8 identical straws make an octagon.

Write a division problem where a 2-digit number of straws are made into octagons and the remainder needs to be **rounded up** in the answer.

Write a second problem where the remainder needs to be **rounded down** in the answer.

Guided practice

Solve these word problems. Decide whether the quotient needs to be rounded up or down.

a Arun uses 5 grams of icing sugar to ice a cake. A bag of icing sugar contains 87 grams. How many cakes can Ryan ice?

Solution: $87 \div 5 = 17 \text{ r } 2$.
2 grams of sugar does not make a full 5 grams needed to ice a cake. Therefore, you round the answer down.
17 r 2 becomes 17, so 17 cakes can be iced.

b A school needs to buy rulers for 89 learners. Rulers are sold in packs of 3. How many packs must the school buy?

$89 \div 3 = 29 \text{ r } 2$
The remaining 2 learners still need a ruler. Therefore, you round the answer up. 29 r 2 becomes 30, so 30 packs are needed.

Lesson 1: **Working with place value (1)**

Key words
- **dividend**
- **divisor**
- **quotient**
- **place value**
- **chunking method**

- Use place value to divide numbers to 1000 by 1-digit numbers

Let's learn

An octopus has 8 tentacles.

How many octopuses together will have 456 tentacles?

Use grouping: How many groups of 8 make 456?

If you know that five 8s make 40 then you also know that fifty 8s make 400.

Now you only need to count from 400 to 456 in 8s to find the answer.

50×8 7×8

$456 \div 8 = 57$

0   400   456

Use the distributive property to break apart the dividend into addends that are multiples of the divisor. Individually divide each addend by the divisor.

Partition 456 into $400 + 56$: $456 \div 8 = \boxed{400 + 56} \div 8$

Divide each addend by 8: $= \boxed{400 + 8} + \boxed{56 \div 8}$

Complete the divisions and find the answer: $= 50 + 7$

57 octopuses will have 456 tentacles.

How many 3-digit divided by 1-digit calculations can you find that give a quotient of 46?

Guided practice

Write two multiplication facts to help find the answer.

$48 \div 4$

$\boxed{10} \times 4 = \boxed{40}$

$\boxed{2} \times 4 = \boxed{8}$

So $48 \div 4 = 12 = \boxed{12}$

Lesson 2: **Working with place value (2)**

Key words
* **expanded written method**
* **chunking method**

* Use place value to divide numbers to 1000 by 1-digit numbers

Let's learn

$288 \div 8 =$ Estimate by rounding: $320 \div 8 = 40$

Step 1

100s	10s	1s

Step 2

100s	10s	1s

3 groups of 80
(or **30** groups of 8)

Step 3

100s	10s	1s

Regroup four 10s as forty 1s.

Step 4

100s	10s	1s

6 groups of 8

$288 = \textbf{30}$ groups of $8 + \textbf{6}$ groups of 8
$= 36$ groups of 8

$288 \div 8 = 36$

Use your 100s, 10s and 1s place value chart and place value counters to work out $235 \div 3$. Discuss each step with your partner.

Guided practice

Fill in the missing numbers and find the quotient.

$172 \div 4 = \boxed{160 \div 4} + \boxed{12 \div 4} = \boxed{40 \div 3} = \boxed{43}$

39

Lesson 3: **Expanded written method**

Key words
- dividend
- quotient
- divisor
- place value
- chunking method

- Use the expanded written method to divide numbers to 1000 by 1-digit numbers

Let's learn

The expanded written method for division involves repeated subtraction of multiples or 'chunks' of the divisor.

Step 1

```
      8
4 ) 3 5 9
```

Step 2

```
      8
4 ) 3 5 9
    3 2 0    4 × 80
```

Step 3

```
      8
4 ) 3 5 9
  − 3 2 0    4 × 80
      3 9
```

Step 4

```
      8 9
4 ) 3 5 9
  − 3 2 0    4 × 80
      3 9
```

Step 5

```
      8 9
4 ) 3 5 9
  − 3 2 0    4 × 80
      3 9
      3 6    4 × 9
```

Step 6

```
      8 9 r 3
4 ) 3 5 9
  − 3 2 0    4 × 80
      3 9
  −   3 6    4 × 9
         3
```

Can you pack 576 tennis balls into containers evenly, so that each container has exactly (with no remainder):

    a 3 balls?   **b** 4 balls?   **c** 5 balls?   **d** 6 balls?

Guided practice

Estimate first, then use the expanded written method of division to work out the answer to the calculation.

$427 ÷ 6 =$ | 71 r 1 |

Estimate: | $420 ÷ 6 = 70$ |

Lesson 4: **Real-life problems**

* Solve problems involving division of numbers to 1000 by 1-digit numbers

Let's learn

For a division calculation, you usually just state the answer and any remainder.

In some word problems, you need to **round** the answer.

Example:

Mia wants to use 787 beads to make bracelets of equal length.

Each bracelet should have 9 beads, if possible.

How many bracelets can she make?

How many bracelets will have 9 beads?

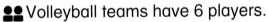

$787 \div 9 = 87 \text{ r } 4$

Mia still needs to use the remaining 4 beads to make a bracelet.

Therefore, she can make 88 bracelets (**rounding up** the remainder) but only 87 of them will have 9 beads (**rounding down** the remainder).

Volleyball teams have 6 players.

Write a division problem in which a 3-digit number of players form teams and the remainder needs to be **rounded up** in the answer.

Write a second problem where the remainder needs to be **rounded down** in the answer.

Guided practice

Decide whether the quotient needs to be rounded up or down.

a A team of athletes joins together to run a 258-kilometre route. Each runner plans to complete a distance of 7 kilometres. How many people will need to run?

$258 \div 7 = 36 \text{ r } 6$

If you round the answer down to 36 runners, then the whole distance will not be covered. Therefore, you need 37 runners (including an extra person required to run the last 6 kilometres).

b How many tickets costing $8 can I buy with $459?

$459 \div 8 = 57 \text{ r } 3$

As tickets cost $8, I can't buy a ticket with the remaining amount. Therefore, I round the answer down to 57 tickets.

Lesson 1: **Decimal place value**

Key words
- decimal
- decimal point
- tenths
- hundredths
- place value

- Explain the value of the tenths and hundredths digits in decimals

Let's learn

Decimal numbers are made up of whole numbers and fractions of numbers.

A dot, called a **decimal point**, separates the whole number from the fraction.

Tenths

The number line between 0 and 1 can be divided into 10 equal parts.

Each of these 10 equal parts is $\frac{1}{10}$. We say: 'one tenth'.

You can write any fraction in tenths as a decimal.

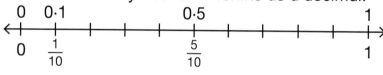

Hundredths

Each interval of $\frac{1}{10}$ can be divided into 10 equal parts.

Each of these 10 equal parts is $\frac{1}{100}$. We say: 'one hundredth'.

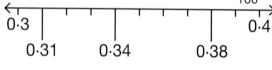

What is the value of each digit in these amounts?

1·83 kg

UNLEADED
2.98

$19.99

Guided practice

What is the value of the underlined digits?

1̲7·6̲8̲

The value of the 1 is 1 ten or 10.

The value of the 8 is 8 hundredths or 0·08.

1 ten
10

10s | 1s • | $\frac{1}{10}$s | $\frac{1}{100}$s
1 | 7 • | 6 | 8

8 hundredths
0·08

7 ones
7

6 tenths
0·6

Lesson 2: **Composing and decomposing decimals**

• Compose and decompose decimals

Let's learn

Decomposing a number means splitting a number into separate smaller parts.
One way to do this is to split a number into its separate place values.

36·79

$$36·79 = 30 + 6 + 0·7 + 0·09$$

Composing a number means forming a number from separate, smaller parts.
One example of this is forming a number from separate place values.

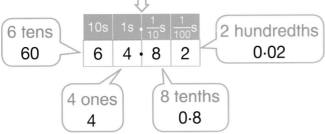

64·82 is composed of:

10s	1s	$\frac{1}{10}$s	$\frac{1}{100}$s
6	4 ·	8	2

6 tens
60

2 hundredths
0·02

4 ones
4

8 tenths
0·8

👥 A number is decomposed into
81 tenths and 9 hundredths.
What is the number?

Guided practice

Decompose the number 82·74 by place value.

$82·74 = 80 + 2 + 0·7 + 0·04$

Lesson 3: **Regrouping decimals**

• Regroup decimals to help with calculations

Let's learn

Numbers can be split by place value but they can also be **decomposed** in other ways.

758 hundredths
7·58

7 ones + 58 hundredths
7 + 0·58

7·58

75 tenths + 8 hundredths
7·5 + 0·08

5 tenths + 708 hundredths
0·5 + 7·08

2 tens + 16 ones + 24 hundredths
20 + 16 + 0·24

6 tens + 624 hundredths
30 + 6·24

36·24

36 ones + 24 hundredths
36 + 0·24

3 tens + 62 tenths + 4 hundredths
30 + 6·2 + 0·04

Write each number as a decimal.

a 6 ones and 37 hundredths

b 31 ones, 6 tenths and 2 hundredths

c 487 hundredths

Guided practice

What is the decimal number that has 67 tenths and 5 hundredths?

As 60 tenths make 6 ones, I know 67 tenths can be written as 6 ones and 7 tenths (6·7).

Including the 5 hundredths, I know the number must be 6·75.

Lesson 4: **Comparing and ordering decimals**

Key words
- compare
- order
- place value

- Compare and order decimals

Let's learn

Comparing decimals

- Compare the digits in place value columns from left to right. So start by comparing whole numbers.
- If the whole numbers are the same, then compare the values of the digits in the tenths place.
- The tenth with the greater value indicates the larger decimal number.
- Continue comparing until all the decimals are ordered.

Order the numbers, greatest to smallest:

6·3, 6·8, 7·1, 5·4, 6·2

10s	1s	$\frac{1}{10}$s
	6 ·	③
	6 ·	⑧
	7 ·	1
	5 ·	4
	6 ·	②

8 tenths is greater than 3 tenths, and 3 tenths is greater than 2 tenths.

7·1 > 6·8 > 6·3 > 6·2 > 5·4

Five children took part in a high-jump competition.
The heights in metres that they jumped are given in the table.

Jacob	Alexa	Hassan	Maria	Neena
1·5	1·2	0·8	1·6	1·1

Who jumped highest?

Who jumped the least high?

Who finished in middle position?

Guided practice

Write the correct symbol, < or >, to compare each pair of decimals.

a 0·8 | > | 0·7

b 6·4 | < | 6·8

Number

Lesson 1: **Multiplying and dividing by 10, 100 and 1000**

- Multiply and divide whole numbers by 10, 100 and 1000

Let's learn

10 000s	1000s	100s	10s	1s	
			4	7	
		4	7	0	× 10
	4	7	0	0	× 100
4	7	0	0	0	× 1000

When multiplied by 10, a number becomes 10 times larger and the digits move 1 place to the left.

When multiplied by 100, a number becomes 100 times larger and the digits move 2 places to the left.

When multiplied by 1000, a number becomes 1000 times larger and the digits move 3 places to the left.

When divided by 10, a number becomes 10 times smaller and the digits move 1 place to the right.

When divided by 100, a number becomes 100 times smaller and the digits move 2 places to the right.

When divided by 1000, a number becomes 1000 times smaller and the digits move 3 places to the right.

	100s	10s	1s	.	$\frac{1}{10}$s	$\frac{1}{100}$s
	8	3	0			
÷ 10		8	3			
÷ 100			8	.	3	
÷ 1000			0	.	8	3

👥 How many different calculations can you write that will give these answers?
1 The calculations should involve multiplying or dividing by a power of 10.

    a 95   **b** 670   **c** 2·62

Use your knowledge of place value.

Guided practice

Complete the calculations.

34 × 10 = 340   34 × 100 = 3400   34 × 1000 = 34 000

610 ÷ 10 = 61   610 ÷ 100 = 6·1   610 ÷ 1000 = 0·61

Lesson 2: **Multiplying decimals by 10 and 100**

Key words
* **multiply**
* **decimal**
* **place value**

* Multiply decimals by 10 and 100

Let's learn

Just as for whole numbers, the digits in a decimal number move 1 or 2 places to the left when multiplied by 10 or 100.

100s	10s	1s	$\frac{1}{10}$s	$\frac{1}{100}$s	
		0 ·	7	5	
		7 ·	5		× 10
	7	5			× 100

When multiplied by 10, a number becomes 10 times larger and the digits move 1 place to the left.

100s	10s	1s	$\frac{1}{10}$s	$\frac{1}{100}$s	
		1 ·	4		
	1	4 ·			× 10
1	4	0			× 100

When multiplied by 100, a number becomes 100 times larger and the digits move 2 places to the left.

100s	10s	1s	$\frac{1}{10}$s	$\frac{1}{100}$s	
		2 ·	8	3	
	2	8 ·	3		× 10
2	8	3			× 100

1 How many different calculations that involve multiplying by 10 or 100 can you write that will give these answers?

    a 23   **b** 7   **c** 506

Use your knowledge of place value.

Guided practice

a How many millimetres is 3·7 cm?

  1 cm = 10 mm   So, 3·7 cm = 3·7 × 10 mm = 37 mm

b How many centimetres is 0·48 m?

  1 m = 100 cm   So, 0·48 m = 0·48 × 100 cm = 48 cm

Number

Lesson 3: **Dividing by 10 and 100**

- Divide decimals by 10 and whole numbers by 10 and 100

Let's learn

Similar to whole numbers, the digits in a decimal number move 1 place to the right when divided by 10.

When divided by 10, a number becomes 10 times smaller and the digits move 1 place to the right.

10s	1s	$\frac{1}{10}$s	$\frac{1}{100}$s
	1 · 7		
÷ 10	0 · 1	7	

10s	1s	$\frac{1}{10}$s	$\frac{1}{100}$s
	5 · 4		
÷ 10	0 · 5	4	

10s	1s	$\frac{1}{10}$s	$\frac{1}{100}$s
4	3 · 5		
÷ 10	4 · 3	5	

The digits in a number move 2 places to the right when divided by 100.

When divided by 100, a number becomes 100 times smaller and the digit moves 2 places to the right.

100s	10s	1s	$\frac{1}{10}$s	$\frac{1}{100}$s
2	3	7		
÷ 100		2 ·	3	7

How many different calculations that involve dividing by 100 can you write that will give these answers?

a 0·66

b 0·08

c 7·91

Use your knowledge of place value.

Guided practice
How many centimetres is 8·3 millimetres?
10 mm = 1 cm   So, 8·3 mm = 8·3 ÷ 10 cm = 0·83 cm

Lesson 4: **Rounding decimals to the nearest whole number**

Number

• Round decimals to the nearest whole number

Let's learn

To round decimals to the nearest whole number, look at the digit in the tenths position.

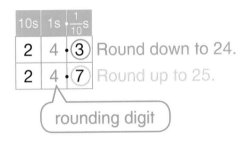

Round down to 24.
Round up to 25.

rounding digit

• If its value is less than 5: the ones digit remains the same.
• If its value is 5 or greater: round up the ones digit to the next whole number.

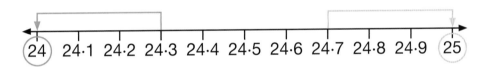

24 24·1 24·2 24·3 24·4 24·5 24·6 24·7 24·8 24·9 25

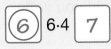 Write all the decimal numbers with one decimal place that round to 7 when rounded to the nearest whole number.

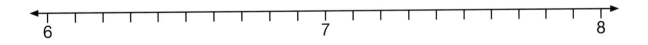

6   7   8

Guided practice

Write the two whole numbers on either side of each decimal.
Then draw a ring around the number that the decimal rounds to.

⑥ 6·4 ⬜7

49

Number

Lesson 1: **Fractions as division**

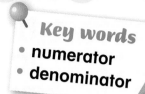
• Understand that a fraction can be represented by a division of the numerator by the denominator

Let's learn

One way to think of a fraction is as a division.

A fraction represents a division of the numerator by the denominator.

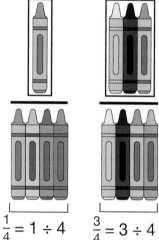

$\frac{1}{4} = 1 \div 4$ $\frac{3}{4} = 3 \div 4$

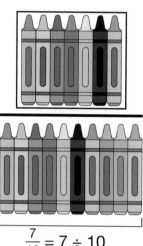

$\frac{7}{10} = 7 \div 10$

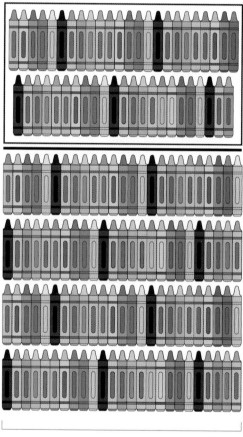

$\frac{47}{100} = 47 \div 100$

$\frac{1}{4} = 1 \div 4$

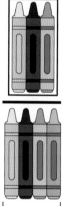

👥 Joel divides a bowl of cherries into four equal parts. What fraction of cherries is in each part? How would you represent this fraction as a division?

Guided practice

Saideep sticks three-tenths of his football stickers in an album. Write this amount of stickers as a fraction of his sticker collection. How would you write this fraction as a division?

I know three-tenths is written as $\frac{3}{10}$.

I can also write this fraction as a division $3 \div 10$.

Lesson 2: **Improper fractions and mixed numbers**

Key words
- numerator
- denominator
- improper fraction
- mixed number

- Recognise improper fractions and mixed numbers

Let's learn

> A **mixed number** is a number that is made up of a whole number and a fraction.

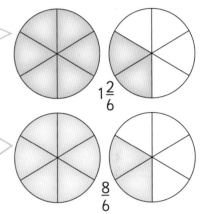

$1\frac{2}{6}$

> In an **improper fraction**, the numerator (top number) is greater than, or equal to, the denominator (bottom number).

$\frac{8}{6}$

Think about your age in years and months.

How would you write this number as a mixed number?

How would you write it as an improper fraction?

 I am $9\frac{2}{3}$ years old

 I am $9\frac{1}{2}$ years old

 I am $10\frac{1}{4}$ years old

Guided practice

Sort the numbers into proper fractions, improper fractions and mixed numbers.

$\frac{4}{3}$ $2\frac{1}{4}$ $\frac{11}{4}$ $\frac{2}{7}$ $1\frac{1}{3}$ $\frac{8}{7}$ $\frac{1}{3}$ $\frac{3}{4}$ $16\frac{2}{5}$

Proper fraction	Improper fraction	Mixed number
$\frac{1}{3}$ $\frac{3}{4}$ $\frac{2}{7}$	$\frac{4}{3}$ $\frac{11}{4}$ $\frac{8}{7}$	$1\frac{1}{3}$ $2\frac{1}{4}$ $16\frac{2}{5}$

Number

Lesson 3: **Converting improper fractions and mixed numbers**

- Convert between improper fractions and mixed numbers

Let's learn

A visual model can help convert improper fractions to mixed numbers.

Here is a circle model for the fraction $\frac{13}{3}$.

Think visually!

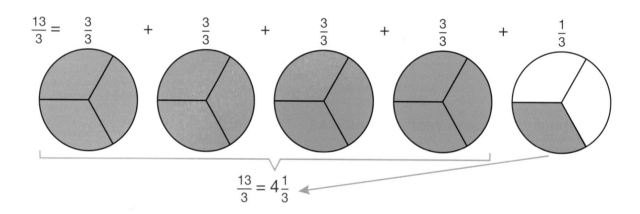

$\frac{13}{3} =$ $\frac{3}{3}$ $+$ $\frac{3}{3}$ $+$ $\frac{3}{3}$ $+$ $\frac{3}{3}$ $+$ $\frac{1}{3}$

$$\frac{13}{3} = 4\frac{1}{3}$$

👥 Hassan said that he and his friends ate $1\frac{1}{4}$ pizzas.

His friend George said that they ate $\frac{5}{4}$ pizzas.

Are they in agreement? How do you know?

Guided practice

Convert $2\frac{3}{4}$ to an improper fraction.

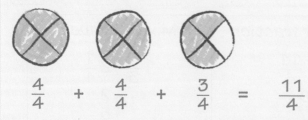

$\frac{4}{4}$ $+$ $\frac{4}{4}$ $+$ $\frac{3}{4}$ $=$ $\frac{11}{4}$

Lesson 4: **Comparing and ordering fractions**

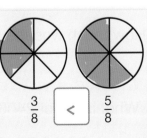

Key words
- numerator
- denominator
- compare
- order

- Compare and order fractions with the same denominator

Let's learn

You can use a fraction wall or number line to compare and order fractions.

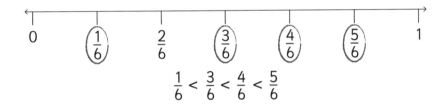

| $\frac{1}{6}$ | $\frac{1}{6}$ | $\frac{1}{6}$ | $\frac{1}{6}$ | $\frac{1}{6}$ | $\frac{1}{6}$ |

| $\frac{1}{6}$ | $\frac{1}{6}$ | $\frac{1}{6}$ | $\frac{1}{6}$ | $\frac{1}{6}$ | $\frac{1}{6}$ |

$$\frac{5}{6} > \frac{4}{6}$$

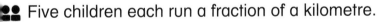

0   $\frac{1}{6}$ $\frac{2}{6}$ $\frac{3}{6}$ $\frac{4}{6}$ $\frac{5}{6}$ 1

$$\frac{1}{6} < \frac{3}{6} < \frac{4}{6} < \frac{5}{6}$$

Five children each run a fraction of a kilometre.

Put the children in order of the distance run, from shortest to longest.

Name	Charlie	Elana	Toby	Farida	Zoe
Distance (km)	$\frac{4}{10}$	$\frac{7}{10}$	$\frac{2}{10}$	$\frac{5}{10}$	$\frac{6}{10}$

Guided practice

Shade each circle to represent the fraction below it.

Then use < or > to compare the fractions.

I shade three parts of the first circle to represent $\frac{3}{8}$.

I shade five parts of the second circle to represent $\frac{5}{8}$.

There are more parts shaded in the second than in the first.

This tells me that $\frac{3}{8} < \frac{5}{8}$.

$\frac{3}{8}$ < $\frac{5}{8}$

Number

Lesson 1: **Fractions as operators**

Key words
- **numerator**
- **denominator**
- **operator**

- Understand that proper fractions can act as operators

Let's learn

To find a non-unit fraction of a quantity:

$\frac{9}{10}$ of $80 =

1 Find one part (unit fraction) by **dividing** the quantity by the denominator.

$80 ÷ 10 = $8

2 Find the required number of parts by **multiplying** the result by the numerator.

$8 × 9 = $72

Sometimes, you can also find a non-unit fraction of a quantity by:

1 Multiplying the quantity by the numerator.

$80 × 9 = $720

2 Then **dividing** the result by the denominator.

$720 ÷ 10 = $72

Lydia says:

I've just won $47!

It's a tenth share of prize money for a competition that I entered!

How much was the total prize?

Guided practice

What is $\frac{1}{100}$ of 2400 ml?

$$\frac{1}{100} \text{ of } 2400 = 2400 ÷ 100$$
$$= 24$$

$\frac{1}{100}$ of 2400 ml = 24 ml

Lesson 2: **Adding and subtracting fractions (1)**

Key words
- numerator
- denominator
- like fractions

Number

- Add and subtract fractions with the same denominator

Let's learn

When adding and subtracting like fractions, add or subtract the numerators, not the denominators. The denominators stay as they are.

> **Remember! Like fractions** are fractions with the same denominator.

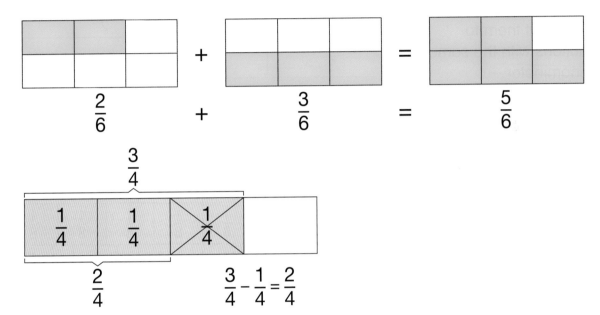

$$\frac{2}{6} + \frac{3}{6} = \frac{5}{6}$$

$$\frac{3}{4}$$

$$\frac{2}{4} \qquad \frac{3}{4} - \frac{1}{4} = \frac{2}{4}$$

👥 How many different ways can you find to balance this number sentence?
1 Choose an example and check if it balances the number sentence.

$$\frac{5}{10} + \frac{\boxed{}}{10} = \frac{8}{10} + \frac{\boxed{}}{10}$$

Guided practice

Use the diagrams to add the fractions.
Shade the last part of the diagram.
Then complete the number sentence.

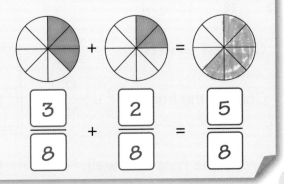

$$\frac{3}{8} + \frac{2}{8} = \frac{5}{8}$$

55

Number

Lesson 3: **Adding and subtracting fractions (2)**

• Add and subtract fractions with different denominators

Let's learn

Unlike fractions are fractions with different denominators. To add or subtract them, you need to convert them to fractions with the same denominator.

To do this, you look for equivalent fractions.

Example: $\frac{3}{4} - \frac{5}{12} =$

Remember to estimate first. $\frac{5}{12}$ is close to $\frac{6}{12}$ ($\frac{1}{2}$ or $\frac{2}{4}$) so $\frac{3}{4} - \frac{5}{12}$ will be close to $\frac{3}{4} - \frac{2}{4} = \frac{1}{4}$. From the fraction wall, you can see that:

$\frac{3}{4} = \frac{9}{12}$

$\frac{3}{4} - \frac{5}{12} = \frac{9}{12} - \frac{5}{12} = \frac{4}{12}$

		1		
$\frac{1}{2}$			$\frac{1}{2}$	
$\frac{1}{3}$		$\frac{1}{3}$		$\frac{1}{3}$
$\frac{1}{4}$	$\frac{1}{4}$		$\frac{1}{4}$	$\frac{1}{4}$
$\frac{1}{5}$	$\frac{1}{5}$	$\frac{1}{5}$	$\frac{1}{5}$	$\frac{1}{5}$
$\frac{1}{6}$	$\frac{1}{6}$	$\frac{1}{6}$	$\frac{1}{6}$	$\frac{1}{6}$ $\frac{1}{6}$
$\frac{1}{8}$ $\frac{1}{8}$	$\frac{1}{8}$ $\frac{1}{8}$	$\frac{1}{8}$	$\frac{1}{8}$	$\frac{1}{8}$ $\frac{1}{8}$
$\frac{1}{9}$ $\frac{1}{9}$	$\frac{1}{9}$ $\frac{1}{9}$	$\frac{1}{9}$	$\frac{1}{9}$ $\frac{1}{9}$	$\frac{1}{9}$ $\frac{1}{9}$
$\frac{1}{10}$ $\frac{1}{10}$	$\frac{1}{10}$ $\frac{1}{10}$	$\frac{1}{10}$ $\frac{1}{10}$	$\frac{1}{10}$ $\frac{1}{10}$	$\frac{1}{10}$ $\frac{1}{10}$
$\frac{1}{12}$ $\frac{1}{12}$	$\frac{1}{12}$ $\frac{1}{12}$	$\frac{1}{12}$ $\frac{1}{12}$	$\frac{1}{12}$ $\frac{1}{12}$ $\frac{1}{12}$	$\frac{1}{12}$ $\frac{1}{12}$ $\frac{1}{12}$

👥 Find two missing numbers that balance the number sentence.

◁1

$$\frac{2}{3} + \frac{\boxed{}}{12} = \frac{16}{12} + \frac{\boxed{}}{4}$$

Guided practice

Convert the fraction: $\frac{3}{5} = \frac{\boxed{6}}{10}$

$\frac{1}{5}$	$\frac{1}{5}$	$\frac{1}{5}$	$\frac{1}{5}$	$\frac{1}{5}$
$\frac{1}{10}$ $\frac{1}{10}$	$\frac{1}{10}$ $\frac{1}{10}$	$\frac{1}{10}$ $\frac{1}{10}$	$\frac{1}{10}$ $\frac{1}{10}$	$\frac{1}{10}$ $\frac{1}{10}$

From the fraction wall, I can see that $\frac{3}{5} = \frac{6}{10}$.

Lesson 4: **Multiplying and dividing unit fractions**

Key words
- unit fraction
- multiply
- divide
- area model

- Multiply and divide unit fractions

Let's learn

Use models to multiply and divide unit fractions by whole numbers.

$\frac{1}{6} \times 4 =$

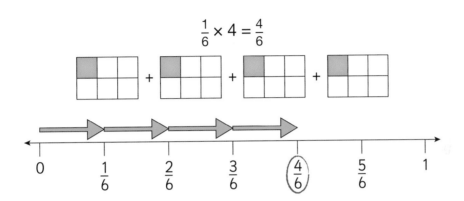

$$\frac{1}{6} \times 4 = \frac{4}{6}$$

$\frac{1}{4} \div 3 =$

1			
$\frac{1}{4}$	$\frac{1}{4}$	$\frac{1}{4}$	$\frac{1}{4}$

$\frac{1}{12}$	$\frac{1}{12}$	$\frac{1}{12}$	$\frac{1}{12}$	$\frac{1}{12}$	$\frac{1}{12}$	$\frac{1}{12}$	$\frac{1}{12}$	$\frac{1}{12}$	$\frac{1}{12}$	$\frac{1}{12}$	$\frac{1}{12}$

Prove that:

4 **a** 3 pieces that are each $\frac{1}{8}$ of pizza is equivalent to $\frac{3}{8}$ of the whole pizza.

b $\frac{1}{3}$ of a chocolate bar shared equally between 3 people will mean that each person gets $\frac{1}{9}$ of the whole bar.

Guided practice

Use the model to multiply.

$\frac{1}{5} \times 3 = \dfrac{\boxed{3}}{\boxed{5}}$

$\frac{1}{5} \times 3 = \frac{1}{5} + \frac{1}{5} + \frac{1}{5} = \frac{3}{5}$

Lesson 1: **Percentages of shapes**

• Recognise percentages of shapes

Key words
• percentage
• denominator

Let's learn

> Remember! **Percentages** tell you the number of parts **in every hundred**. The sign for 'per cent' is %.

In the picture, there are 100 cones.

Three cones are red and the rest are blue.

To describe the red cones, you could say three-hundredths ($\frac{3}{100}$) or you could express it as a percentage: 3 per cent, or 3%.

 What percentage of each shape is shaded?

a

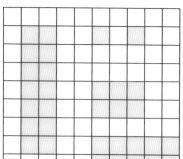

b

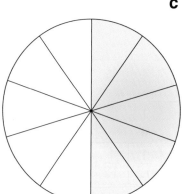

c

Guided practice

Shade the squares in the grid to show 73%.

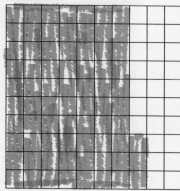

73% means 73 out of 100.

To show 73%, I shade 73 out of the 100 squares on the grid.

58

Key words
- percentage
- denominator

Lesson 2: **Converting between fractions and percentages**

• Write percentages as a fraction with denominator 100

Let's learn

A **percentage** can be thought of as another name for **hundredths**.

A fraction expressed as a hundredth can be renamed as a percentage.

For example: $\frac{1}{100} = 1\%$ $\frac{27}{100} = 27\%$

$\frac{6}{10}$ of these pencils are blue. This is the same as $\frac{60}{100}$.

So, you can say that 60% of these pencils are blue.

Harry has 100 football stickers. He has stuck 63 of the stickers in an album.

What percentage of Harry's stickers are not in the album?

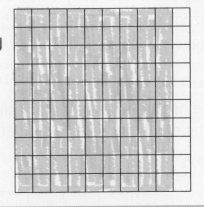

Guided practice

Asma says:

9 out of 10 of my friends walk to school.

Shade the squares on the grid to show this fraction and then convert it to a percentage.

I can represent '9 out of 10' by shading 9 out of 10 parts, 90 out of 100 squares, on the grid.

$\frac{9}{10} = \frac{90}{100} = 90\%$

Number

Number

Lesson 3: **Comparing percentages**

Key words
* percentage
* quantity
* compare

* Compare percentages and tenths

Let's learn

$70\% = \frac{7}{10}$

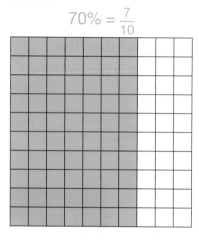

$40\% = \frac{4}{10}$

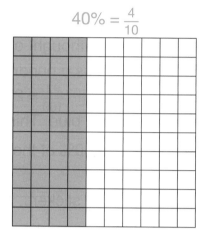

$\frac{7}{10} > \frac{4}{10}$

or

$70\% > 40\%$

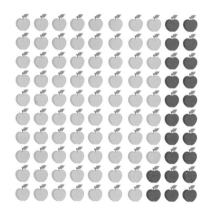

$78\% < 87\%$

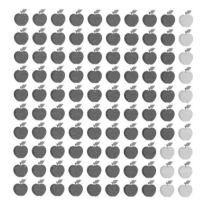

Zina eats $\frac{4}{5}$ of a pizza. Lucy eats 75% of a pizza.

If both pizzas are the same size, who eats more pizza?

Explain how you know.

Guided practice

Mark the numbers on the number line. Draw a ring around the greater number and complete the number statement with the symbol < or >.

$\frac{1}{2}$ | < | 60%

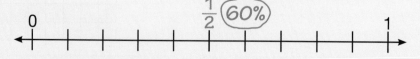

$\frac{1}{2}$ ⟨60%⟩

0   1

Lesson 4: **Ordering percentages**

Key words
• percentage
• quantity
• order

• Order percentages and tenths

Let's learn

$60\% = \frac{6}{10}$ $50\% = \frac{1}{2}$ $40\% = \frac{4}{10}$

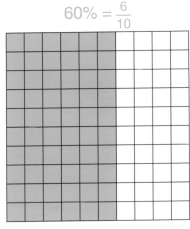 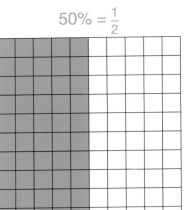 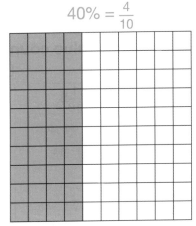

$60\% > 50\% > 40\%$ or $60\% > \frac{1}{2} > 40\%$

Maisie, Zikra and Jamie are each given identical cakes.

Maisie eats 70% of her cake. Zikra eats $\frac{3}{5}$ of her cake.
Jamie eats 50% of his cake.

Who has eaten the greatest amount of cake?

Who has eaten the smallest amount of cake?

Explain how you know.

Guided practice

Mark the numbers on the number line. Then write the numbers in order from smallest to greatest.

30%, $\frac{1}{4}$, 20%

$\boxed{20\%} < \boxed{\dfrac{1}{4}} < \boxed{30\%}$

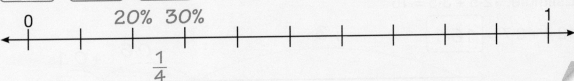

0   20% 30%   1

$\frac{1}{4}$

Lesson 1: **Adding decimals (1)**

• Add pairs of decimals mentally

Let's learn

You can use mental strategies to add decimals. Remember to estimate first.
3.7 is close to 4 so $5.5 + 3.7$ will be close to $5.5 + 4 = 9.5$

$5.5 + 3.7 =$

Using place value counters

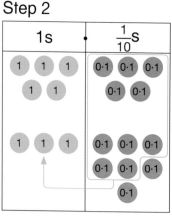

Using a number line (counting on)

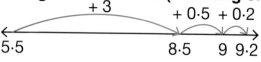

leading to

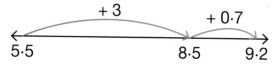

4·7 litres   2·8 litres

👥 How many litres are there in total?

Which method did you use to work out the answer?

Guided practice

Add the numbers mentally. Use any strategy you prefer. Remember to estimate first.

Estimate: $12.5 + 3.5 = 16$

$12.4 + 3.7 =$ | 16·1 |

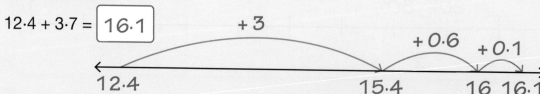

Lesson 2: **Adding decimals (2)**

Key words
• **place value**
• **regroup**
• **addend**

Number

• Add pairs of decimals using written methods

Let's learn

You can use written methods to add decimals.

Partitioning

$5.7 + 7.6 =$

Partition both numbers and find the sum.

$$5.7 + 7.6 = 5 + 0.7 + 7 + 0.6$$
$$= 5 + 7 + 0.7 + 0.6$$
$$= 12 + 1.3$$
$$= 13.3$$

Partition one number and find the sum.

$$5.7 + 7.6 = 5.7 + 7 + 0.6$$
$$= 12.7 + 0.6$$
$$= 13.3$$

Remember to estimate first.
$28.86 + 36.77$ is close to
$29 + 37$ so the answer will
be close to 66.

Formal written method

$28.86 + 36.77 =$

$$\begin{array}{r} 28.86 \\ + \ 36.77 \\ \hline 65.63 \\ \hline {\scriptstyle 1 \ \ 1 \ \ 1} \end{array}$$

I use partitioning
when numbers are small
or have few digits.

I use the formal
written method when
numbers get bigger.

● Eddie, Aisha and Magnus run on Monday and Tuesday.

2 They complete these distances.

Eddie: 18·74 km and 19·67 km

Aisha: 13·63 km and 24·78 km

Magnus: 22·47 km and 15·94 km

Find the total distance run for each person. Remember to estimate first.
Look at the results. What do you notice?

Guided practice

Solve by partitioning one number. Remember to estimate first.

Estimate: $5.5 + 2.5 = 8$

$5.6 + 2.7 = \boxed{8.3}$

$$5.6 + 2.7 = 5.6 + 2 + 0.7$$
$$= 7.6 + 0.7$$
$$= 8.3$$

Lesson 3: **Subtracting decimals (1)**

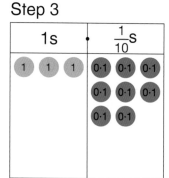

Key words
• place value
• regroup

• Subtract pairs of decimals mentally

Let's learn

You can use mental strategies to subtract decimals. Remember to estimate first. 2·7 is close to 2·5 so 6·5 – 2·7 will be close to 6·5 – 2·5 = 4.

6·5 – 2·7 =

Using place value counters

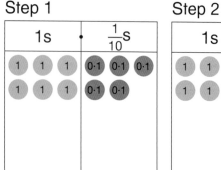 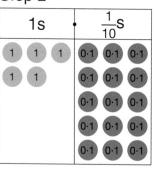

Using a number line (counting back)

leading to

👥 Hassan has $8.63 and spends $4.26.

How much money does he have left?

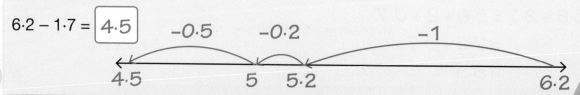

Guided practice

Subtract the numbers mentally. Use any strategy you prefer. Remember to estimate first.

Estimate: 6 – 1·5 = 4·5

6·2 – 1·7 = 4·5

–0·5   –0·2   –1

4·5   5   5·2   6·2

Lesson 4: **Subtracting decimals (2)**

Key words
* place value
* regroup
* subtrahend

* Subtract pairs of decimals using written methods

Let's learn

You can use written methods to subtract decimals.

$8·3 − 4·7 =$

Partition the subtrahend and complete the subtraction

$$8·3 − 4·7 = 8·3 − 4 − 0·7$$
$$= 4·3 − 0·7$$
$$= 3·6$$

Use a known number fact

$$83 − 47 = 36$$
$$8·3 − 4·7 = 3·6 \text{ (10 times smaller than } 83 − 47 = 36)$$

Formal written method

$74·56 − 56·77 =$

```
   6 13 14 1
   7̶4̶·5̶6̶
 − 56·77
   17·79
```

Four friends go shopping. Find the missing amounts in the table.

	Money to spend	Money spent	Money left
Hilda	$43.64	$18.78	
Max	$67.32		$28.65
Clara	$56.53	$27.66	
Caspar		$57.67	$26.78

Guided practice

Solve by partitioning the subtrahend. Remember to estimate first.

Estimate: $6·5 − 4 = 2·5$

$6·3 − 3·8 = \boxed{2·5}$

$$6·3 − 3·8 = 6·3 − 3 − 0·8$$
$$= 3·3 − 0·8$$
$$= 2·5$$

Number

Lesson 1: **Working with place value**

- Use place value to multiply numbers with one decimal place by 1-digit whole numbers

Let's learn

$5.6 \times 4 =$

×	5	0·6
4	1 1 1 1 1 / 1 1 1 1 1 / 1 1 1 1 1 / 1 1 1 1 1	0·1 0·1 0·1 0·1 0·1 0·1 / 0·1 0·1 0·1 0·1 0·1 0·1 / 0·1 0·1 0·1 0·1 0·1 0·1 / 0·1 0·1 0·1 0·1 0·1 0·1

$5.6 \times 4 = \boxed{5 \times 4} + \boxed{0.6 \times 4}$

$= 20 + 2.4$

$= 22.4$

$46.3 \times 3 =$

×	40	6	0·3
3	10 10 10 10 / 10 10 10 10 / 10 10 10 10	1 1 1 1 1 1 / 1 1 1 1 1 1 / 1 1 1 1 1 1	0·1 0·1 0·1 / 0·1 0·1 0·1 / 0·1 0·1 0·1

$46.63 \times 3 = \boxed{40 \times 3} + \boxed{6 \times 3} + \boxed{0.3 \times 3}$

$= 120 + 18 + 0.9$

$= 138.9$

👥 Sang uses place value counters to multiply 8·3 by 6. How many ones counters will he need? How many tenths counters? What is the product?

Guided practice

Draw a diagram to show the calculation. Then use it to find the product.

$3.6 \times 4 = \boxed{14.4}$

×	3	0·6
4	1 1 1 / 1 1 1 / 1 1 1 / 1 1 1	0·1 0·1 0·1 0·1 0·1 0·1 / 0·1 0·1 0·1 0·1 0·1 0·1 / 0·1 0·1 0·1 0·1 0·1 0·1 / 0·1 0·1 0·1 0·1 0·1 0·1
Total	12	2·4

Lesson 2: **Using times table facts**

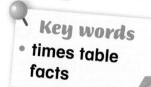

Number

• Multiply numbers with one decimal place by 1-digit whole numbers, using times table facts and number lines

Let's learn

$0.6 \times 7 =$

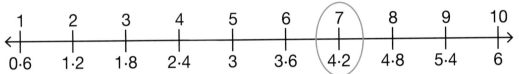

From the number line, you can see that $0.6 \times 7 = 4.2$.

What is $0.8 \times 9 =$

Since $8 \times 9 = 72$, you know that $0.8 \times 9 = 7.2$ (10 times smaller).

You can also use a table of facts.

×	0·7	0·8	0·9	1
7	4·9	5·6	6·3	7
8	5·6	6·4	7·2	8
9	6·3	7·2	8·1	9
10	7	8	9	10

Predict which of these calculations gives the smallest/greatest product before you calculate.

Use the strategy you prefer to calculate each product.

Remember to estimate the answers first.

a $0.6 \times 8 \times 3 =$ **b** $0.7 \times 7 \times 5 =$ **c** $0.8 \times 8 \times 4 =$ **d** $0.7 \times 9 \times 6 =$

Guided practice

Label the number line in steps of 0·8 to answer the questions.

$0.8 \times 5 = \boxed{4}$ $0.8 \times 8 = \boxed{6.4}$

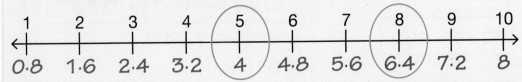

Lesson 3: **Grid method**

- Multiply numbers with one decimal place by 1-digit whole numbers, using the grid method

Key words
- grid method
- partition
- partial product

Let's learn

You can multiply decimal numbers using the grid method in a similar way to multiplying whole numbers.

Whole numbers

$249 \times 3 =$

Estimate by rounding: $250 \times 3 = 750$

$\boxed{200 \times 3}$ + $\boxed{40 \times 3}$ + $\boxed{9 \times 3}$

×	200	40	9
3	600	120	27

$600 + 120 + 27 = 747$

$249 \times 3 = 747$

Decimal numbers

$24.9 \times 3 =$

Estimate by rounding: $25 \times 3 = 75$

$\boxed{20 \times 3}$ + $\boxed{4 \times 3}$ + $\boxed{0.9 \times 3}$

×	20	4	0.9
3	60	12	2.7

$60 + 12 + 2.7 = 74.7$

$24.9 \times 3 = 74.7$

Use these cards. Make four calculations involving numbers less than 100 with one decimal place multiplied by a one-digit number. For example, 75.1×3.

$\boxed{\cdot}$ $\boxed{1}$ $\boxed{3}$ $\boxed{5}$ $\boxed{6}$ $\boxed{7}$ $\boxed{8}$

Conjecture and predict which of the calculations will give the greatest product and which will give the smallest product.
How do you know?

Use the grid method to find the product of each calculation.

Order the calculations, from smallest to greatest product.

Guided practice

Estimate first, then use partitioning to work out the answer. Show your working. Check your answer with your estimate.

$73.6 \times 7 = \boxed{515.2}$

Estimate: $\boxed{70 \times 7 = 490}$

$\boxed{70 \times 7}$ + $\boxed{3 \times 7}$ + $\boxed{0.6 \times 7}$

$= 490 + 21 + 4.2$

$= 515.2$

Lesson 4: **Written methods**

- Multiply numbers with one decimal place by 1-digit whole numbers, using a written method

Number

Let's learn

$8.8 \times 6 =$

Estimate by rounding: $9 \times 6 = 54$

Partitioning

$$8.8 \times 6 =$$

$8 \times 6 \qquad 0.8 \times 6$

$4.8 \quad + \quad 48 \quad = 52.8$

Expanded written method

8.8×6 is equivalent to $88 \times 6 \div 10$

```
        10s  1s
          8   8
     ×        6
          4   8     8 × 6
      4   8   0     80 × 6
      5   2   8
          1
```

$528 \div 10 = 52.8$

Calculate the area of each field.

Use any written strategy.

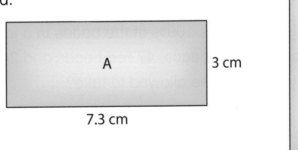

A   3 cm

7.3 cm

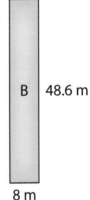

B   48.6 m

8 m

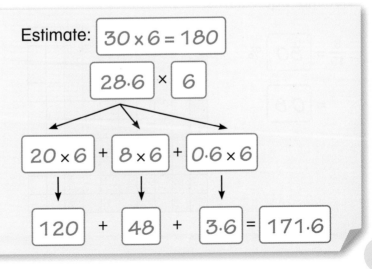

Guided practice

Use partitioning to multiply.
Estimate the answer first.

$28.6 \times 6 =$ | 171.6 |

Estimate: | $30 \times 6 = 180$ |

| 28.6 | × | 6 |

| 20 × 6 | + | 8 × 6 | + | 0.6 × 6 |

| 120 | + | 48 | + | 3.6 | = | 171.6 |

Number

Lesson 1: **Fraction, decimal and percentage equivalence (1)**

• Know that proper fractions, decimals and percentages can have equivalent values

Let's learn

Fractions, decimals and percentages are all linked. We can express one as any of the others.

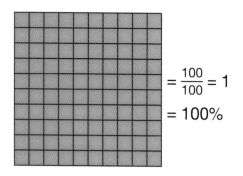

$= \frac{100}{100} = 1$

$= 100\%$

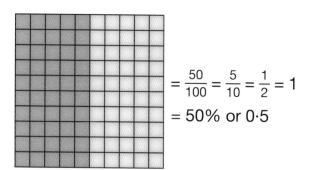

$= \frac{50}{100} = \frac{5}{10} = \frac{1}{2} = 1$

$= 50\%$ or 0.5

Noor is allowed to take 100% of the beads in a jar.

The jar has 18 green beads, 47 red beads and 24 blue beads.

How many beads is she allowed to take?

Guided practice

Shade the fraction on the 100 grid and write the percentage and decimal equivalents.

$\frac{8}{10} = \boxed{80}$ %

$= \boxed{0.8}$

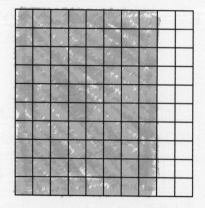

70

Lesson 2: **Fraction, decimal and percentage equivalence (2)**

Key words
- proper fraction
- decimal
- percentage
- equivalent

- Know that proper fractions, decimals and percentages can have equivalent values

Number

Let's learn

Remember! Decimals and percentages are just different ways of describing fractions.

Part of a whole				
Fraction	$\frac{20}{100} = \frac{2}{10} = \frac{1}{5}$	$\frac{40}{100} = \frac{4}{10} = \frac{2}{5}$	$\frac{50}{100} = \frac{5}{10} = \frac{1}{2}$	$\frac{70}{100} = \frac{7}{10}$
Percentage	20%	40%	50%	70%
Decimal	0·2	0·4	0·5	0·7

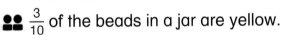 $\frac{3}{10}$ of the beads in a jar are yellow.

0·4 of the beads are orange.

The remainder are purple.

Write the number of purple beads as a percentage of the total number of beads.

Guided practice

Shade the fraction on the 100 grid and write the percentage and decimal equivalents.

$\frac{80}{100} = \boxed{80}$ %

$= \boxed{0·8}$

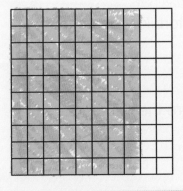

Number

Lesson 3: Comparing fractions, decimals and percentages

- Compare numbers with one decimal place, proper fractions and percentages, using the symbols =, > and <

Let's learn

> **Remember!** Fractions, decimals and percentages can all be used to describe the same part of a whole or set.

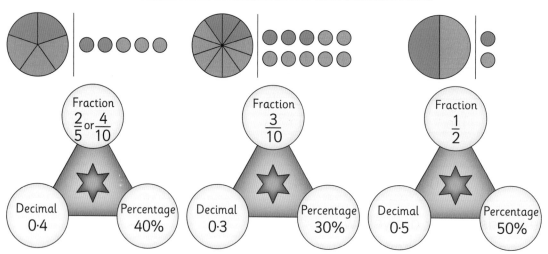

To compare fractions, decimals and percentages, it is often easier to write them in the same form.

Which is greater, 50% or 0·4?

$0·4 = 40\%$ so $50\% > 0·4$

Which is smaller, $\frac{2}{5}$ or 30%?

$\frac{2}{5} = 40\%$ so $30\% < \frac{2}{5}$

👥 In a shopping centre, $\frac{1}{2}$ of the vehicles parked are cars.

40% of the vehicles parked are motorbikes.

Compare the numbers of parked motorbikes and cars. Which is greater?

Guided practice

Use the symbols <, > or = to compare the numbers.

$\frac{7}{10}$ ⟶ > ⟶ 60%

Lesson 4: **Ordering fractions, decimals and percentages**

Key words
* proper fraction
* decimal
* percentage

• Order numbers with one decimal place, proper fractions and percentages, using the symbols =, > and <

Let's learn

Remember! Fractions, decimals and percentages are just different ways of expressing the same value.

Converting fractions, decimals and percentages into the same form can help us to compare and order them.

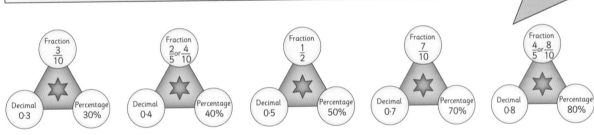

Ascending order

Fraction $\frac{3}{10}$	Fraction $\frac{2}{5}$ or $\frac{4}{10}$	Fraction $\frac{1}{2}$	Fraction $\frac{7}{10}$	Fraction $\frac{4}{5}$ or $\frac{8}{10}$
Decimal 0·3 / Percentage 30%	Decimal 0·4 / Percentage 40%	Decimal 0·5 / Percentage 50%	Decimal 0·7 / Percentage 70%	Decimal 0·8 / Percentage 80%

George places five trays of shapes on a table.
The trays are labelled A to E. Each tray contains 100 shapes.

Tray A: 60% are triangles. Tray B: $\frac{3}{10}$ are squares.

Tray C: $\frac{1}{2}$ are circles. Tray D: $\frac{4}{10}$ are rectangles.

0·3 of the shapes in Tray E are hexagons.

Put the shapes in ascending order of their numbers in the tray.

Guided practice

Order the numbers: 80%, $\frac{2}{5}$, 0·1, $\frac{1}{5}$, 30%

First convert all of the numbers to decimals.

80%	$\frac{2}{5}$	0·1	$\frac{1}{5}$	30%
0.8	0.4	0.1	0.2	0.3

Write the decimals in ascending order.

0.1   0.2   0.3   0.4   0.8

$$\boxed{0.1} < \boxed{\frac{1}{5}} < \boxed{30\%} < \boxed{\frac{2}{5}} < \boxed{80\%}$$

73

Number

Lesson 1: **Proportion (1)**

• Understand that proportion compares part to whole

Let's learn

A proportion compares a part of a whole with the whole amount.

We describe a proportion using the terms 'in every' or 'out of'.

We often write proportion as a fraction.

Each small box of chocolates contains 1 white chocolate and 3 dark chocolates.

Proportion of dark:
3 in every 4.
3 out of 4.
$\frac{3}{4}$ are dark.

Proportion of white:
1 in every 4.
1 out of 4.
$\frac{1}{4}$ are white.

Mateo has a jar of sweets.

3 in every 8 sweets are cherry-flavoured.
5 in every 8 sweets are mint-flavoured.

What proportion of sweets are mint-flavoured?

What is the smallest number of sweets that could be in the jar?

Guided practice

Complete the sentences.

| 1 | in every | 3 | flowers is red.

The fraction of flowers that are red is $\dfrac{1}{3}$

| 2 | in every | 3 | flowers are yellow.

The fraction of flowers that are yellow is $\dfrac{2}{3}$

Lesson 2: **Proportion (2)**

• Describe proportions using fractions and percentages

Number

Let's learn

Remember! A proportion tells us how much we
have of something compared to the whole amount.

We can write a proportion as a fraction.

3 in every 5 beads are red.
$\frac{3}{5}$ are red.

2 in every 5 beads are blue.
$\frac{2}{5}$ are blue.

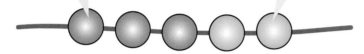

If 2 beads in every 5 beads are blue, then it follows that there will be:

• 4 blue beads in 10 beads

• 6 blue beads in 15 beads… and so on.

We can also express a proportion as a percentage.

60% of the beads are red.

40% of the beads are blue.

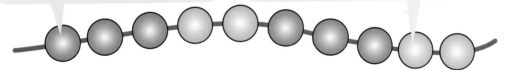

In every 10 tins of paint, 8 tins are large and 2 tins are small.

As a percentage, what proportion of the tins are large?

What proportion are small?

Guided practice

In every plate of muffins, there are 3 blueberry muffins and 7 banana muffins.
Write the proportion of blueberry muffins on a plate as a fraction and as a
percentage.

Do the same for the banana muffins.

There are 10 muffins on every plate: 3 + 7 = 10

Proportion of blueberry = $\frac{3}{10}$ = 30%

Proportion of banana = $\frac{7}{10}$ = 70%

Number

Lesson 3: **Ratio**

Key words
• ratio
• for every

• Understand that ratio compares part to part of two or more quantities

Let's learn

A **ratio** compares two or more numbers or quantities. It shows how much larger or smaller one part is than the others.

We describe a ratio using the term 'for every'.

We write a ratio using the symbol :

We place this symbol between the quantities we are comparing.

> These cubes are in the ratio 2 red to 1 blue.
> There are 2 red cubes for every 1 blue cube.
> The ratio of red cubes to blue cubes is 2:1.

We can also say that:

> These cubes are in the ratio 1 blue to 2 red.
> There is 1 blue cube for every 2 red cubes.
> The ratio of blue cubes to red cubes is 1:2.

Rio has a bag of red and green sweets. There are 3 green sweets for every 4 red sweets.

Rio writes the ratio of green to red sweets as 4:3.

Is this correct? If not, explain why.

Guided practice

Write the ratio of black circles to white circles.

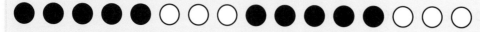

Black to white: ⬛ 5 : 3

The ratio is 5 black circles for every 3 white circles.

Lesson 4: **Ratio and proportion**

- Use ratio and proportion to make comparisons

Number

Let's learn

> **Remember!** A **proportion** tells us how much we have of something compared to the whole amount.
>
> A **ratio** tells us how much we have of one amount compared to another amount.

The ratio of squares to circles to triangles is 3:5:2.

The proportion of squares is $\frac{3}{10}$ or 30%.

The proportion of circles is $\frac{5}{10}$, $\frac{1}{2}$ or 50%.

The proportion of triangles is $\frac{2}{10}$, $\frac{1}{5}$ or 20%.

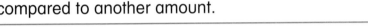

The ratio of blue to green to red shapes is 4:1:5.

The proportion of blue shapes is $\frac{4}{10}$, $\frac{2}{5}$ or 40%.

The proportion of green shapes is $\frac{1}{10}$ or 10%.

The proportion of red shapes is $\frac{5}{10}$, $\frac{1}{2}$ or 50%.

What is the ratio of circles to squares?

What is the proportion of each type of shape?

What is the ratio of red shapes to blue shapes to green shapes?

What is the proportion of each coloured shape?

Guided practice

Write the ratio and proportion of black to white circles.

 Ratio 1 : 9

Proportion of black circles

$\frac{1}{10}$ or 10 %

Proportion of white circles

$\frac{9}{10}$ or 90 %

Lesson 1: **Calculating time intervals (1)**

- Understand time intervals of less than one second

Geometry and Measure

Let's learn

Here are some things that take less than one second.

Your heart to beat once

A sneeze

Your eye to blink

A skydiver to fall 30 m

A wiper to cross the windscreen

A clap

Use a stopwatch to time these events.

6

| Say the alphabet from A to E | Laugh | Turn around a full circle | Count to 10 |

| Write the letter 'S | Walk to the door from the centre of the classroom | Jump up and down 5 times | Read a short sentence |

Draw a chart and classify them into events that take 'less than one second' and events that take 'longer than one second'.

Guided practice

Logan used a stopwatch to time four events. The results are in the table.

Event	a clap	a jump on the spot	a blink	to count 1, 2, 3
Duration (time taken)	0·23 s	0·6 s	0·2 s	0·58 s

Order the times taken for the events, from the shortest to the longest.

The order is 0·2s < 0·23s < 0·58s < 0·6s.

Lesson 2: **Calculating time intervals (2)**

Key words
- **time interval**
- **duration**

- Find time intervals in seconds, minutes and hours

Let's learn

Tom catches the 6:35 p.m. bus arriving home at 7:05 p.m.
How long is his journey?

Method 1:

Count in 5- and 10-minute intervals.

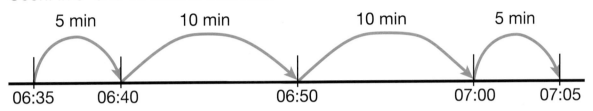

| 5 min | 10 min | 10 min | 5 min |

06:35   06:40   06:50   07:00   07:05

Method 2: Say:

> 6:35 and 25 minutes takes you to 7:00, so add another 5 minutes.

Total: 30 minutes

Janina looks at a train timetable and thinks she spots some errors.

Is Janina correct?

If so, help her to correct the mistakes.

Start time	End time	Elapsed time
3:55 a.m.	7:23 a.m.	2 hours 28 minutes
11:50 p.m.	2:26 a.m.	2 hours 36 minutes
9:25 a.m.	11:51 a.m.	2 hours 16 minutes
4:57 p.m.	6:41 p.m.	2 hours 44 minutes
11:31 p.m.	**2:51 a.m.**	3 hours 20 minutes

Guided practice

Tom boards a train at 4:35 p.m. He arrives at his destination at 6:10 p.m.
How long does his journey take?

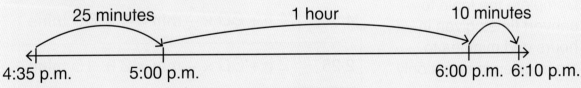

| 25 minutes | 1 hour | 10 minutes |

4:35 p.m.   5:00 p.m.   6:00 p.m. 6:10 p.m.

25 min + 1 hour + 10 min = 1 hour 35 min

Lesson 3: **Expressing time intervals**

Key words
- time interval
- duration

- Express time intervals as decimals, or in mixed units

Let's learn

Time is not metric, so you need to be careful when converting from one unit to another.

60 seconds is 1 minute.

60 minutes is 1 hour.

24 hours is 1 day.

7 days is 1 week.

365 days, 52 weeks or 12 months is 1 year.

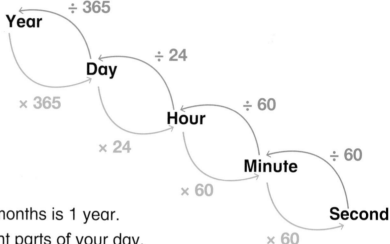

Year ÷ 365 Day ÷ 24 Hour ÷ 60 Minute ÷ 60 Second
× 365 × 24 × 60 × 60

👥 Think about the different parts of your day.

- Getting ready for school (washing, brushing teeth, breakfast, packing school bag)
- Travelling to school
- The school day: lessons, lunch, breaks
- After school/the evening: watching TV, doing homework, dinner, getting ready for bed

Create a timetable of your day. Write the approximate times for each activity or event, for example: 'breakfast – 15 minutes' or 'morning lessons – 2 hours 45 minutes'.

Next, convert these times to decimals or mixed units, for example, 15 mins = 0·25 hour, 2 hours 45 mins = 2·75 hours.

Write the converted times in brackets alongside the original units.

Guided practice

Complete the table to convert times given in hours and minutes to times given in hours only.

Hours	Minutes
4·5	$4\,h + 60 \times \frac{1}{2}\,min = 4\,h,\ 30\,mins$
2·25	$2\,h + 60 \times \frac{1}{4}\,min = 2\,h,\ 15\,mins$
3·75	$3\,h + 60 \times \frac{3}{4}\,min = 3\,h,\ 45\,mins$

Lesson 4: **Compare times between time zones**

• Compare times between different time zones

Let's learn

The world is divided into 24 different time zones. Clocks are set one hour apart in adjacent time zones.

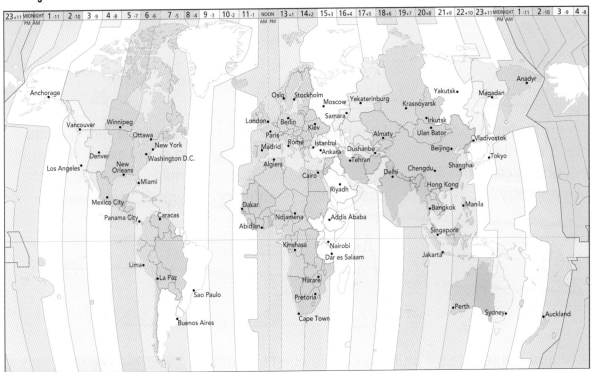

Look at the time zone map.

Have you visited any of these countries? Choose one country and use the map to work out what time it is there.

If not, choose any country and work out the time.

What time will it be in that country when the time here is 6 a.m., 6 p.m., 9 p.m.?

Guided practice

Jun lives in New York. His uncle lives in Beijing.

It is 10 a.m. in New York. What time is it in Beijing?

The time zone for New York is −5 and for Beijing is +8.

So, New York and Beijing are 13 (5 + 8) time zones apart.

The time in Beijing is 11 p.m. (10 a.m. + 13 hours).

Geometry and Measure

Lesson 1: Identifying, describing and classifying triangles

Key words
- equilateral
- isosceles
- scalene
- acute angle
- obtuse angle
- right angle
- reflective symmetry

- Identify, describe and classify triangles

Let's learn

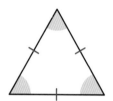

An **equilateral** triangle has 3 equal sides, 3 equal angles and 3 lines of symmetry.

An **isosceles** triangle has 2 equal sides, 2 equal angles and at least 1 line of symmetry.

A **scalene** triangle has no equal sides, no equal angles and no lines of symmetry.

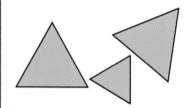

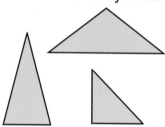

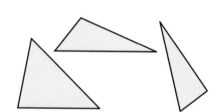

A shape **tessellates** if multiple copies can fit together to cover a flat surface without leaving any gaps.

Investigate and characterise which of the three types of triangle will tessellate.

Which type(s) of triangle can be used to form a regular octagon?

Will a regular octagon tessellate?

Guided practice
Name the triangle. Explain how you know.

This is an isosceles triangle.

It is an isosceles triangles because it has two equal angles and two equal sides.

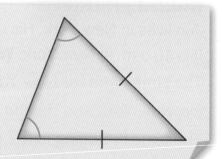

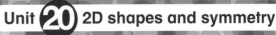

Lesson 2: **Sketching triangles**

- Describe and draw the three types of triangle

Let's learn

It is easy to draw an isosceles triangle on squared paper.

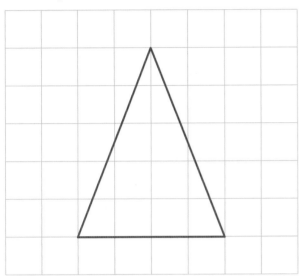

Draw an isosceles triangle on squared paper. Make the base 5 units and the height 3 units.

Draw a scalene triangle.

Is it easy to draw an equilateral triangle on squared paper? If not, why?

Guided practice

Use the triangular dot paper to draw:

a a scalene triangle with two of its sides 4 cm and 3 cm

b an isosceles triangle with a base of 3 cm.

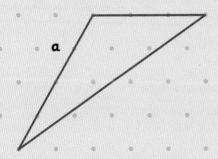
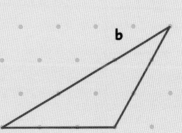

Geometry and Measure

83

Lesson 3: **Identifying symmetrical patterns**

- Identify symmetrical patterns

Key words
- reflective symmetry
- line of symmetry
- mirror line

Let's learn

Some patterns have one line of reflective symmetry. Some have two lines of reflective symmetry, at right angles to each other.

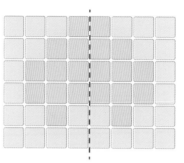 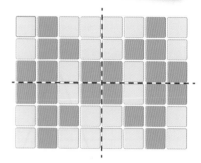

Which of these patterns have reflective symmetry? How do you know?

A 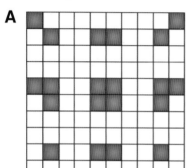   B 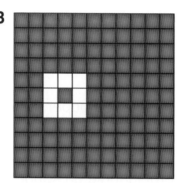   C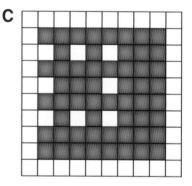

Guided practice

How many lines of symmetry does this pattern have?
Draw the lines of symmetry on the pattern.

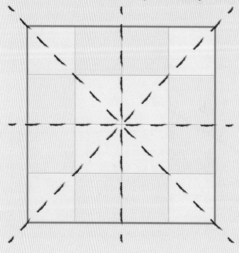

The pattern has four lines of symmetry.

Geometry and Measure

Lesson 4: **Completing symmetrical patterns**

Key words
- reflective symmetry
- line of symmetry
- mirror line

- Create symmetrical patterns

Let's learn

To make a symmetrical pattern, you need to make sure that the pattern on one side of the line of symmetry is exactly the same as the pattern on the other side, but reversed.

Is this pattern symmetrical? Use a mirror to check.

Colour a 3 by 3 square grid to make a symmetrical pattern.

4

Next, copy the pattern three times and place the four grids together to make a bigger square.

Investigate whether the pattern is still symmetrical. Why do you think this is?

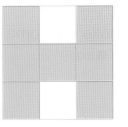

Will this work for grids of any size?

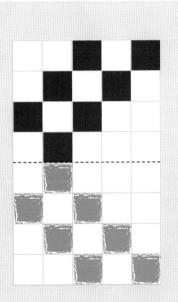

Geometry and Measure

Guided practice

Complete this pattern so that it has one line of symmetry.

Lesson 1: **Identifying and describing 3D shapes**

- Identify and describe 3D shapes

Let's learn

You can identify 3D shapes from their properties, whatever their orientation.

These properties include the number and shape of faces and the number of edges and vertices, if they have them.

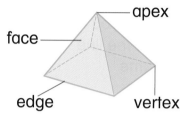

A cube has 6 square faces, 12 edges and 8 vertices. It does not matter which position it is in, it will always be a cube.

cube   cube

👥 Choose a 3D shape to model.
Use straws and modelling clay to make the frame of the shape. Next, cover the frame with plain paper to produce faces.

Once you have made your shape, try positioning it in different orientations. How many different ways can you place your shape in space?

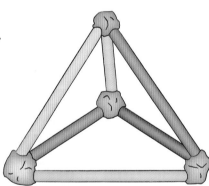

Key words
- prism
- pyramid
- vertex (vertices)
- apex
- polyhedron
- polyhedra

You will need
- 3D shapes
- straws
- modelling clay
- sticky tape
- scissors
- plain paper

Guided practice

One of the faces of a 3D shape has the shape below. What could the shape be?

I know that a pyramid and a triangular prism have triangular faces. It could be either of those shapes.

Lesson 2: **Sketching 3D shapes**

- Describe and sketch 3D shapes

Let's learn

You can use drawings to represent a 3D object on a 2D surface.

You use dashed lines to show the edges and faces you cannot see.

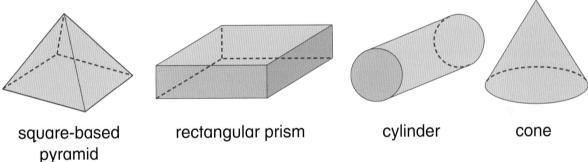

| square-based pyramid | rectangular prism | cylinder | cone |

Think of a 3D shape. Don't tell your partner what it is. Draw the different 2D shapes that are the faces of your 3D shape.

Swap papers with your partner and use the 2D shapes to identify the 3D shape.

Guided practice
What is the name of this shape?

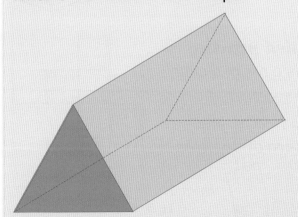

The shape has triangular ends connected by three rectangles.
I think this shape is a triangular prism.

Geometry and Measure

Lesson 3: **Nets for a cube (1)**

• Identify and sketch different nets for a cube

Geometry and Measure

Let's learn

A **net** is a flat shape that can be folded to make a 3D shape.

Nets show which parts are the base, top and sides when folded up.

| net of an open cube | open cube (folded) | net of a closed cube | closed cube (folded) |

There may be several nets for one shape, for example, these are some of the nets of a closed cube.

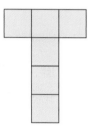 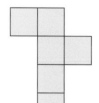 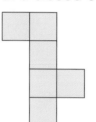 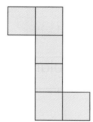 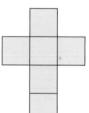 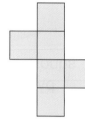

•• Here are six faces
1 of a cube:

Here is the cube viewed from three different angles:

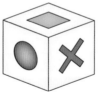  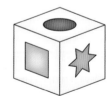

Copy this net onto squared paper and draw the six faces in the correct places.

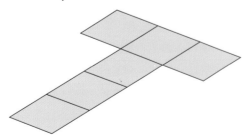

Guided practice

Is this a net of a cube?

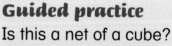

I try to picture in my head folding the pattern of squares.

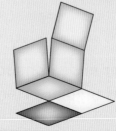

Yes, it does form a cube.

Lesson 4: **Nets for a cube (2)**

- Identify and sketch different nets for a cube

Let's learn

There may be several nets for one shape.
Here are some nets for a closed cube.

There are
11 possible nets for
a closed cube.

Six of the nets
are given. Use
interlocking
squares to find the
remaining five nets.

Sketch the nets
on a piece of
squared paper.

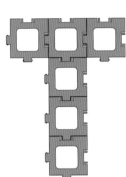 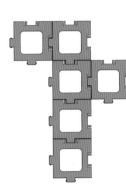 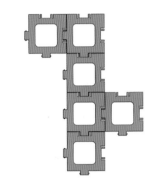

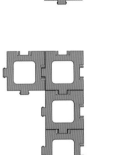 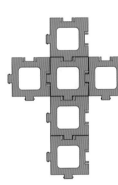 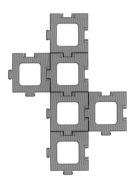

Geometry and Measure

Guided practice

Does this pattern of squares
form a cube?

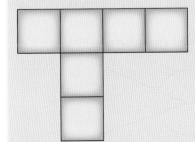

I try to picture in my head
folding the pattern of squares.

No, it does not form a cube. Two
of the square faces overlap.

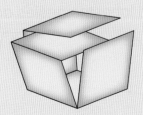

Lesson 1: **Estimating and classifying angles**

• Estimate and classify angles

Key words
• angle
• right angle
• acute angle
• obtuse angle
• straight angle
• reflex angle
• full angle

Let's learn

Right angle
Angle of 90°. Its symbol is a little square box.

Acute angle
Angle less than 90°.

Obtuse angle
Angle greater than 90° but less than 180°.

Reflex angle
Angle greater than 180° but less than 360°.

Which is the best angle to hang the painting? Explain your answer.

If the floor was not perfectly horizontal would this still be the best angle? Explain.

Guided practice

Identify and label the angles acute, right, obtuse, straight or reflex.

A   B   C

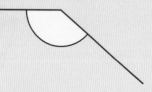

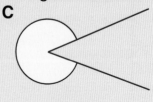

acute   obtuse   reflex

Geometry and Measure

Lesson 2: **Comparing angles**

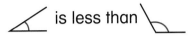
• Compare angles

Let's learn
We can compare angles.

 is less than is greater than

Key words
• angle
• right angle
• acute angle
• obtuse angle
• straight angle
• reflex angle
• full angle

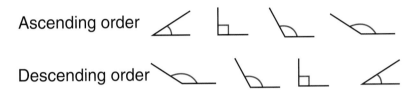

Ascending order

Descending order

👥 You will need four pairs of scissors.

Open each pair so the blades form angles:

• two angles less than a right angle (acute)
• one at a right angle
• one angle greater than a right angle, but less than a straight angle (obtuse).

Arrange the angles in ascending order.

 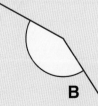 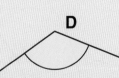

Geometry and Measure

Guided practice
Each angle has a letter. Place the angles in ascending order by writing the letters in the boxes.

C | D | E

A

B

C	<	E	<	A	<	D	<	B

Lesson 3: **Angles on a straight line (1)**

- Calculate unknown angles on a straight line

Geometry and Measure

Let's learn

Angles that meet on a straight line always add up to 180°.

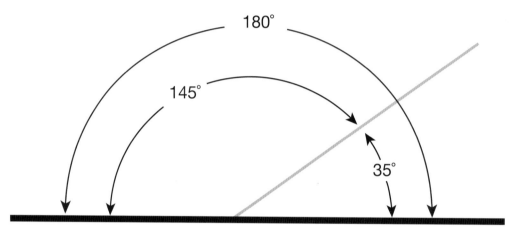

:: Two angles X and Y meet at a point on a straight line.

a If angle X is 137°, what is angle Y?

b If angle X is equal to angle Y, what is angle X?

c If angle Y is three times the size of angle X, what is angle Y?

Guided practice

What is the size of the angle labelled x?

$131° + x = 180°$

$x = 180° - 131°$

$x = 49°$

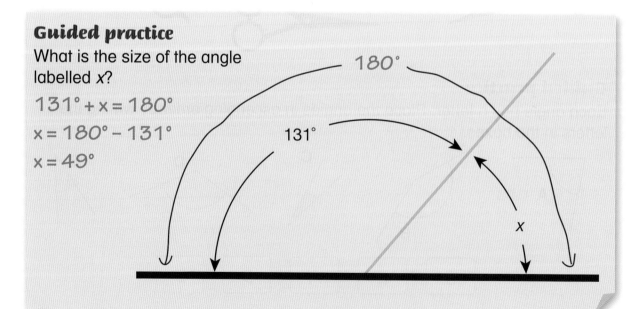

Lesson 4: **Angles on a straight line (2)**

• Calculate unknown angles on a straight line

Let's learn

Finding an unknown angle when more than two angles meet at a point on a straight line is similar to solving a calculation with unknown numbers.

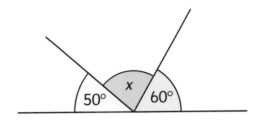

Since angles on a straight line add to 180°:

$50° + x + 60° = 180°$

(simplify the calculation) $\quad 110° + x = 180°$

(using the inverse operation) $\quad x = 180° - 110°$

$= 70°$

Therefore, angle x is 70°.

Three angles X, Y and Z meet at a point on a straight line.

a If angle X is 47° and angle Y is 86°, what is angle Z?

b If all three angles are equal, what is angle Y?

c Angle X is two times the size of angle Y. Angle Z is three times the size of angle Y. What are angles X, Y and Z?

Geometry and Measure

Guided practice
What is the size of the angle labelled x?

$37° + 88° + x = 180°$

$125° + x = 180°$

$x = 180° - 125°$

$x = 55°$

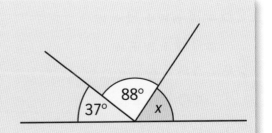

Lesson 1: **Perimeter of simple 2D shapes**

- Measure and calculate the perimeter of simple 2D shapes

Geometry and Measure

Let's learn

You can find the perimeter of a rectangle by any of these methods.
- Measure and find the sum of the sides.
- Double the sum of the length and width. The rule is $P = 2 \times (l + w)$.

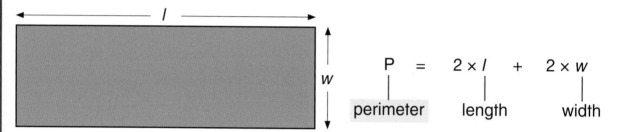

$$P = 2 \times l + 2 \times w$$

perimeter   length   width

The perimeter of a rectangle is 72 centimetres.

4 The length of the rectangle is three times the width of the rectangle.

What is the length of the rectangle? How do you know?
Convince your partner.

Guided practice

Find the width of this rectangle, given its length and perimeter.

I can find the perimeter (P) of a rectangle by doubling the sum of the length (*l*) and the width (*w*).

$28 = 9 + w + 9 + w$

$\quad = 18 + w + w$

$28 - 18 = w + w$

$\quad 10 = w + w$

So, $w = 5$ metres.

The width of the rectangle is 5 m.

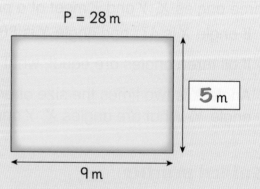

$P = 28$ m

5 m

9 m

My estimate: 9 is close to 10.
Double the width will be close to 28 – double $10 = 8$
Width will be close to 4 m.

Lesson 2: **Perimeter of compound 2D shapes**

- Calculate the perimeter of compound 2D shapes

Let's learn

Cutting the square corner out of a rectangle keeps its perimeter the same.

Rather than adding together all the individual side lengths, you can find the perimeter by thinking of it as a single rectangle.

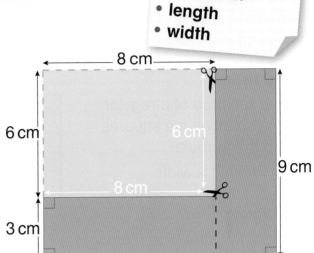

A school playground is a compound shape made of a rectangle 14 metres by 11 metres and a square with sides 12 metres placed side by side. The head teacher wants a fence that will surround the entire edge of the playground. The cost of the fencing is $20 per metre.

What is the total cost of the fencing?

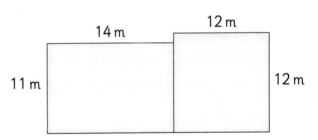

Geometry and Measure

Guided practice

What is the perimeter of the garden lawn?

Think of the lawn as a single rectangle with a 'cut corner'.

$P = 2 \times (l + w)$
$\quad = 2 \times (10 + 11)$
$\quad = 2 \times 21$
$\quad = 42\,m$

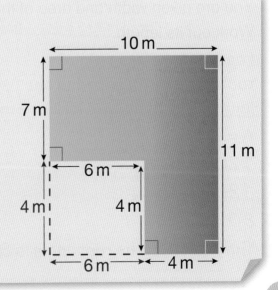

95

Lesson 3: **Area of simple 2D shapes**

- Measure and calculate the area of simple 2D shapes
- Understand that shapes with the same perimeter can have different areas and vice versa

Key words
- area
- length
- width
- square units
- square metre (m²)

Let's learn

To find the area of a regular shape, count all the squares, or use the formula:

Area = length × width

or

$A = l \times w$

length

width

The length of the blue rectangle is 8 units.

The width is 3 units.

Area = length × width

= 8 × 3

= 24 square units

👥 The perimeter of a rectangle is 80 metres and its length is 23 metres. What is the area of the rectangle? How do you know?

Guided practice

You are given width and area of this rectangle.

Work out its length.

I can find the area (A) of a rectangle by multiplying the length (*l*) by the width (*w*).

$A = l \times w$

$1200 = l \times 15$

$1200 \div 15 = l$

$80 = l$

The length of the rectangle is 80 m.

15 m | A = 1200 m²

Lesson 4: **Area of compound 2D shapes**

- Calculate the area of compound 2D shapes

Let's learn

It is sometimes necessary to calculate the area of a compound shape.

You can find the area of rectilinear shapes by splitting them into non-overlapping rectangles and finding the sum of their areas.

Or, you can find the area of a 'missing piece' rectangle and subtract this from the larger rectangle.

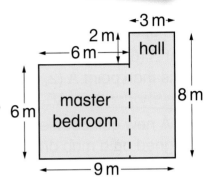

A new wooden floor is to be laid in a kitchen. The room is the shape of an 11 m by 8 m rectangle, with a 'missing piece' in one corner that is the shape of a 3 m by 3 m square.

The cost of the flooring is $25 per square metre.

How much will the new floor cost in total?

Guided practice

What is the area of the shape?

I split the shape into two smaller shapes with a green dotted line.

Area = $(2 \times 4) + (8 \times 8)$

　　　= $8 + 64$

　　　= $72\,m^2$

To check, I calculated the area of the larger rectangle (10 m by 8 m).

Then I subtracted the area of the 'missing piece' (2 m by 4 m).

Area = $(10 \times 8) - (2 \times 4)$

　　　= $80 - 8$

　　　= $72\,m^2$

Lesson 1: **Comparing coordinates (1)**

- Compare two points plotted on the coordinate grid to say which is closer to each axis

Let's learn

Point A (2, 6) is closer to the *y*-axis than point B (6, 3) because AC is shorter than BD.

Point B (6, 3) is closer to the *x*-axis than point A (2, 6) because BF is shorter than AE.

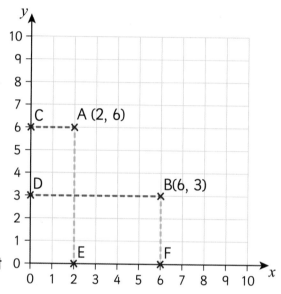

A new golf course is being designed on a map drawn on a coordinate grid.

A compass on the map shows that the *y*-axis points directly north.

Two holes, E and F, are located at coordinates (7, 6) and (3, 9).

The next two holes, G and H, must be located so that:

- hole G is east of hole E
- hole H is south of hole E and west of hole F.

Provide possible coordinates for the locations of holes G and H.

Guided practice

Point S is located at (8, 4).

Point T is located at (3, 7).

Which point is further away from the *x*-axis?

Which point is closer to the *y*-axis?

Point T is further away from the
x-axis than point S.

Point T is closer to the y-axis than
point S.

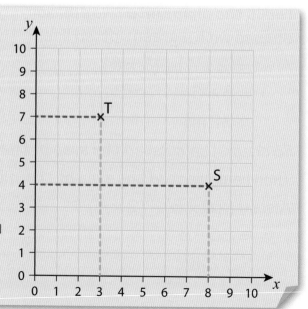

Geometry and Measure

Lesson 2: **Comparing coordinates (2)**

- Without a grid, estimate the position of point A relative to point B given the coordinates of point B

Let's learn

The *x*-coordinate gives the horizontal distance from the *y*-axis.

The *y*-coordinate gives the vertical distance from the *x*-axis.

As the *x*-coordinate of point A is smaller than the *x*-coordinate of point B ($4 < 6$), point A will be closer to the *y*-axis.

As the *y*-coordinate of point A is greater than the *y*-coordinate of point B ($7 > 3$), point A will be further away from the *x*-axis.

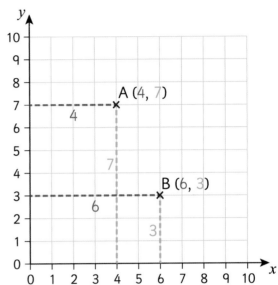

Draw a set of *x*- and *y*-axes on a piece of plain paper. Label each axis from 0 to 10.

Estimate the position of point A (6, 6), which is one vertex of a square with sides of 4 units.

Estimate the positions of the other three vertices, B, C and D.

Guided practice

This is point E (7, 3).

Estimate the position of point F (3, 6).

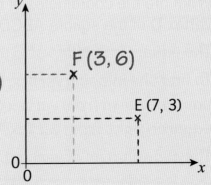

I can see that the *x*-coordinate of point F (3) is smaller than the *x*-coordinate of point E (7), so point F is closer to the *y*-axis.

As 3 is less than half of 7, I estimate the position of point F along the *x*-axis to be slightly less than halfway between the origin and 7.

The *y*-coordinate of point F (6) is greater than the *y*-coordinate of point E (3), so point F is further away from the *x*-axis.

As 6 is double 3, I estimate the position of point F along the *y*-axis to be twice the distance between the origin and 3.

Lesson 3: **Plotting coordinates (1)**

Key words
- *x*-axis
- *y*-axis
- **coordinates**

- Plot points to form squares in the first quadrant

Geometry and Measure

Let's learn

Given A, B and C, the coordinates of three vertices of a square you can use the properties of a square (equal sides) to find the coordinates of the missing vertex D.

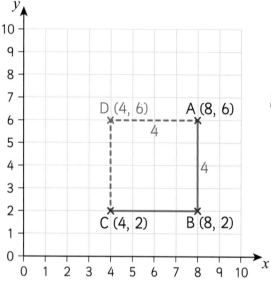

👥 Four boys make a square on a coordinate grid painted on the playground.

Two boys stand at diagonally opposite corners of the square with coordinates (7, 6) and (2, 1).

Where should the other two boys stand? Write the coordinates of these points.

Guided practice

What are the coordinates of the missing vertex D of the square ABCD?

The square has sides of 6 units.

To find the missing vertex, I move 6 squares down from vertex C (or 6 squares right from vertex A).

The coordinates of vertex D are (9, 3).

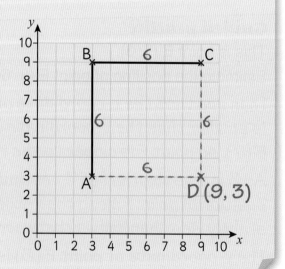

Lesson 4: **Plotting coordinates (2)**

Key words
- *x*-axis
- *y*-axis
- **coordinates**

- Plot points to form shapes in the first quadrant

Let's learn

Given A, B and C, the coordinates of three vertices of a rectangle, you can use the properties of a rectangle (opposite sides equal) to find the missing vertex D.

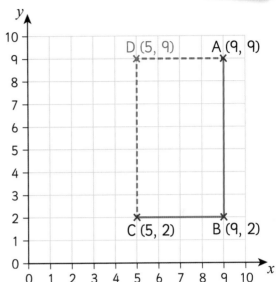

Four girls make a rectangle on a coordinate grid painted on the playground.

Two girls stand at diagonally opposite corners of the rectangle with coordinates (4, 1) and (0, 9).

Where should the other two girls stand? Write the coordinates of these points.

Geometry and Measure

Guided practice

Draw a rectangle. Start at point (3, 4). Write the coordinates of the vertices of the shape.

I decide that my rectangle will be 5 units long and 2 units wide.

I plot a point at (3, 4).

I then move right 5 units and plot a point at (8, 4).

Next, I move down 2 units and plot a point at (8, 2)

Lastly, I move left 5 units and plot a point at (3, 2) and join all the vertices of the rectangle.

The coordinates of the vertices are (3, 4), (8, 4), (8, 2) and (3, 2).

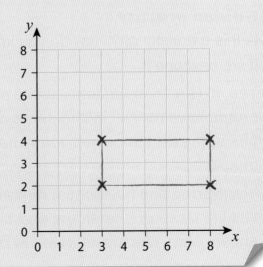

Lesson 1: **Describing translations**

- Describe the translation of a 2D shape on a square grid

Key words
- **translate**
- **translation**
- **orientation**
- **vertex** (vertices)

Let's learn

A **translation** is a movement of a shape in a straight line.

It can be moved up, down, left or right, but you cannot rotate, stretch or change it.

Each vertex must move in the same direction and by the same amount.

The object and its image are identical in shape, size and orientation.

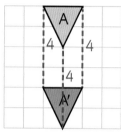

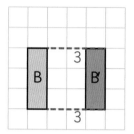

Shape A has been translated 4 squares down.

Shape B has been translated 3 squares to the right.

A table (A) is moved across the classroom. Describe the translation.

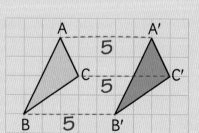

Guided practice

Is this a translation? How do you know?

If so, describe it.

Yes, it is a translation. All the points in the shape have moved in the same direction and the same distance, without changing the shape.

Measuring the horizontal distance between the vertices of the shape and the image shows that the translation is 5 squares right.

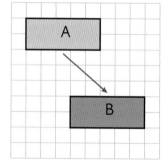

Geometry and Measure

Lesson 2: **Performing translations**

- Perform the translation of a 2D shape on a square grid

Let's learn

Toby is asked to translate triangle ABC 6 squares left and 4 squares up.

To do this, he translates each vertex of the triangle 6 squares to the left and 4 squares up.

The vertices of the image are A', B' and C'.

Toby then joins the vertices of the image to form the triangle in its new translated position.

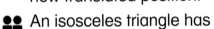

An isosceles triangle has a base of 4 units and a height of 8 units.

Draw the triangle in the centre of a piece of squared paper and translate the shape 7 squares to the left and 6 squares down.

How do you know the translation has been successful?

Guided practice

Translate the shape right 7 squares, down 6 squares.

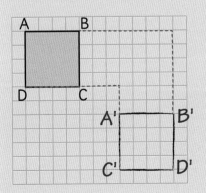

Lesson 3: **Reflecting shapes (1)**

- Reflect 2D shapes in both horizontal and vertical mirror lines on square grids

Key words
- reflection
- object
- image
- line of symmetry
- mirror line
- parallel
- vertex (vertices)

Let's learn

A **reflection** is like 'flipping' an object over a mirror line to produce an image.

Every point on the image is the same distance from the mirror line as the corresponding point on the object.

The image is the same size and shape as the object.

It has not changed. It is just in a different position.

👥 A rectangle is drawn on
4 squared paper.

One pair of sides is parallel with a horizontal mirror line.

Draw the rectangle after reflection.

How do you know your reflection is correct?

Mirror line

4 units	4 units
object 2 units	2 units image
4 units	4 units

Geometry and Measure

Guided practice

Reflect the shape in the horizontal mirror line.
Mark the position of the image.

Work out the vertical distance from each vertex ABCD to the mirror line.

The vertices of the image will be the same distance from the mirror line but on the other side.

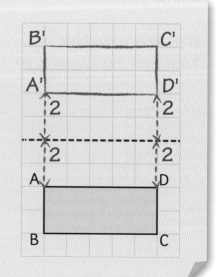

Lesson 4: **Reflecting shapes (2)**

• Predict and draw where a shape will be after reflection where the sides of the shape are not vertical or horizontal

Let's learn

Shapes can have sides that are not parallel or perpendicular. The methods for reflecting these shapes is the same as if they were parallel or perpendicular.

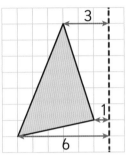 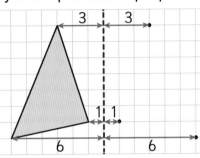 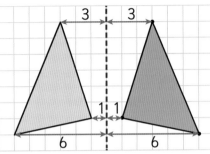

Measure the distance of each vertex from the mirror line (the measurement line must be perpendicular to the mirror line).

Measure the same distance on the opposite side of the mirror line and mark the image of the vertex.

Connect the reflected vertices to form the image of the shape.

👥 Is this a correct reflection of the
4 triangle in the mirror line?

How do you know?

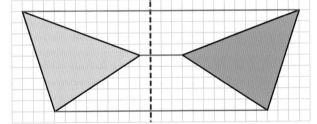

Guided practice

Reflect the rectangle in the mirror line.

I measure from each vertex to the mirror line.

Next, I measure the same distance again on the other side of the mirror line and I mark the image of each vertex.

Then I connect the vertices of the image.

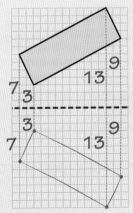

Geometry and Measure

Lesson 1: **Venn and Carroll diagrams**

Key words
• statistical question
• Venn diagram
• Carroll diagram

• Know how to construct a statistical question
• Represent data in Venn and Carroll diagrams

Let's learn

Statistical questions are those that require the collection of data.

Rewrite each question as a statistical question.

a How many siblings does your friend have?

b How many points did your favourite sports team score in its last game?

c How many learners are there in our class?

d Can I do a cartwheel?

What is a statistical question?

Statistical	Not statistical
• What colours do learners in my class prefer?	• Does your dog weigh more than my dog?
• How many hours a day do learners in my class spend playing video games?	• What is Freddy's shoe size?
	• How does Rita get to school?
• How many goals were scored by each football team this season?	

Guided practice

Class 7 collected data to answer the question: *How many learners in our class play netball, football or both sports?*

They recorded the data in a Carroll diagram.

Use the diagram to answer these questions.

	Play football	Don't play football
Play netball	5	9
Don't play netball	11	8

• How many learners play netball but do not play football? $\boxed{9}$

• How many learners play both sports? $\boxed{5}$

• How many learners do not play either sport? $\boxed{8}$

Statistics and Probability

Lesson 2: Tally charts, frequency tables and bar charts

Key words
- statistical question
- tally chart
- frequency table
- bar chart

- Know how to construct a statistical question
- Represent data in tally charts, frequency tables and bar charts

Let's learn

Tally charts and frequency tables are useful for collecting information. Tally charts use marks to record results; frequency tables use numbers.

This table shows data gathered to investigate the statistical question: *Which colour is the most popular with learners in Stage 5?*

Colour	Tally	Frequency
pink	~~IIII~~ III	8
blue	~~IIII~~ ~~IIII~~ IIII	14
yellow	~~IIII~~	5
green	III	3
red	IIII	4
orange	~~IIII~~ IIII	9
purple	~~IIII~~ II	7
	Total	50

Work with a partner to complete your own survey of favourite colours in the class.

Record data in a tally chart. Find the frequency of each colour.

What frequency of learners in your class prefer green?

Which colour was the most popular?

Guided practice

Class 7 collected this data to investigate favourite colours in Stage 5.

What percentage of learners preferred blue or yellow?

38%

Colour	Tally	Frequency	Percentage
pink	~~IIII~~ III	8	16%
blue	~~IIII~~ ~~IIII~~ IIII	14	28%
yellow	~~IIII~~	5	10%
green	III	3	6%
red	IIII	4	8%
orange	~~IIII~~ IIII	9	18%
purple	~~IIII~~ II	7	14%
	Total	50	

Lesson 3: **Waffle diagrams**

- Know how to construct a statistical question
- Represent data in waffle diagrams

Let's learn

A **waffle diagram** is a square 10 by 10 grid. Each box (or **cell**) represents 1%.

It is useful for comparing percentages and showing how each percentage makes up the whole.

How many children prefer a book about space? (53%)

How many children prefer a book about animals? (35%)

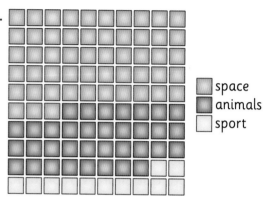

- space
- animals
- sport

How much greater is the percentage of children that prefer a book about animals than a book about sport? (35% − 12% = 23%)

👥 Discuss the waffle diagram in Guided practice with your partner. What conclusions and conjectures can you make from the data?

Guided practice

The table shows the data collected when 100 children were asked to name their favourite pet, from a choice of dog, cat, rabbit or hamster.

Pet	Frequency	Percentage
dog	55	55%
cat	25	25%
rabbit	17	17%
hamster	3	3%

Represent the data in a waffle diagram.

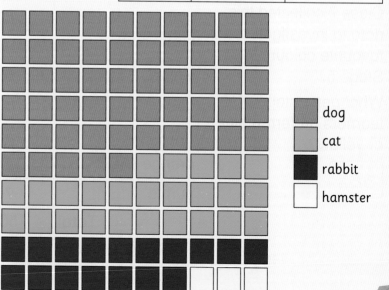

- dog
- cat
- rabbit
- hamster

Lesson 4: **Mode and median**

- Find and interpret the mode and the median of a data set

Key words
- mode
- median

Let's learn

An average is a value that represents the most common number in the set or the middle number in a data set.

Mode is one type of average.

It is the value that appears most frequently in a data set.

A data set may have one mode, more than one mode, or no mode at all.

2, 7, 4, 1, 6, 7, 9
The mode is 7.

30 cm, 45 cm, 17 cm, 30 cm, 45 cm, 30 cm
The mode is 30.

120 g, 80 g, 75 g, 80 g, 110 g, 75 g, 95 g, 80 g, 75 g
The modes are 75 and 80.

Median is another type of average.

It is the middle value in an ordered (ascending or descending) list of values.

Order the set of values. The median is the middle value.
2, 7, 3, 1, 6, 7, 9
1, 2, 3, ⑥, 7, 7, 9
The median is 6.

26, 54, 38, 15, 52, 29, 18
The median is 29.

238 km, 177 km, 218 km, 186 km, 209 km
The median is 209 km.

Write a set of seven data values in which the mode is 4 and the median is 7.

Guided practice

The heights of 7 chicks are measured:
14 cm, 13 cm, 12 cm, 13 cm, 15 cm, 11 cm, 16 cm

What is the mode of their heights? $\boxed{13}$

What is the median? $\boxed{13}$

Statistics and Probability

Lesson 1: **Frequency diagrams and line graphs**

Key words
* **statistical question**
* **frequency diagram**
* **line graph**

* Know how to construct a statistical question
* Represent data in frequency diagrams and line graphs

Let's learn

A **bar chart** displays data that can only take certain values, such as a number of books or a sort of fruit.

A **frequency diagram** displays data that can take on any value, such as height or mass. The values are recorded in groups.

The vertical axis shows the frequency in both cases.

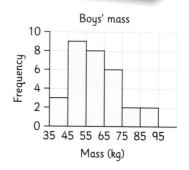

Boys' mass

A **line graph** displays data that changes over time, such as distance walked or how temperature varies.

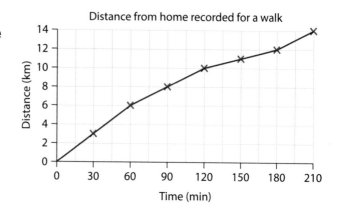

Distance from home recorded for a walk

The line graph shows how far Lucy walked over 3·5 hours.
  a How far had she walked after 75 minutes?
  b In which time interval did she walk the fastest?

Guided practice

Use the frequency diagram in Let's learn to answer these questions.

a How many boys have a mass between 65 kg and 75 kg? | 6 |

b How many boys have a mass between 35 kg and 45 kg? | 3 |

c Between which masses is the mass of most boys? | 45 kg and 55 kg |

d How many boys have a mass more than 75 kg? | 4 |

Statistics and Probability

Lesson 2: **Dot plots**

- Know how to construct a statistical question
- Represent data in dot plots

Let's learn

A **dot plot** is based on a horizontal number line.

The number of dots drawn above a value shows the frequency of that value.

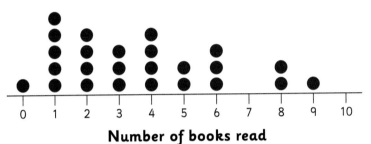

Books read this month

Number of books read

This data shows the total number of goals scored per match over 36 matches.

4	3	2	2	4	3	0	4	2
3	1	4	2	1	5	6	5	1
7	2	0	3	0	3	4	1	2
2	4	6	4	2	2	3	3	0

Draw a dot plot to display the data.

Write three facts about the plot.

Draw a dot plot to display the data.

Write three facts about the plot.

Guided practice

Use the dot plot in Let's learn to answer these questions.

a Which number of books read had the highest frequency? | 1 |

b Which number of books read had the lowest frequency? | 7 and 10 |

c How many learners read more than 7 books? | 3 |

d How many learners were surveyed? | 25 |

Statistics and Probability

111

Lesson 3: **Probability (1)**

- Describe the likelihood of an event happening, using the language of chance

Let's learn

Probability is the study of random events — how likely something is to happen. Understanding probability is essential for interpreting weather forecasts.

MON	TUE	WED	THU	FRI	SAT	SUN
40%	40%	60%	40%	40%	30%	50%

Probability can be shown on a scale.

The terms 'certain', 'likely', 'even chance', 'unlikely' and 'impossible' describe how likely an event is to happen.

unlikely   likely

impossible   even chance   certain

Even chance events have an equally likely chance of happening. For example, rolling a 1 to 6 dice and getting an odd number is equally likely as rolling an even number. This is because there are equal numbers of odd numbers (1, 3, 5) and even numbers (2, 4, 6) on the dice.

Draw and label a likelihood scale.

Label the scale with terms that describe the likelihood of an event happening, such as 'impossible', 'even chance', 'certain'.

Describe probability events below the scale and draw a line to the point on the scale that indicates the likelihood of these events taking place.

Guided practice

Look at the numbers on the spinner. Then tick the boxes that apply.

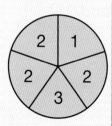

Event	Probability		
	impossible	possible	certain
spinning 4	✓		
spinning 3		✓	
spinning 1 or a prime number			✓
spinning a letter	✓		

Statistics and Probability

Lesson 4: **Probability (2)**

• Describe the results of chance experiments, using the language of probability

Key words
• likelihood
• probability
• chance
• impossible
• unlikely
• even chance
• likely
• certain

Let's learn

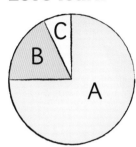

Spinning a letter is **certain**.
Spinning a letter 'D' is **impossible**.
Spinning a letter 'B' is **possible**.
Spinning a letter 'A' is **likely**.
Spinning a letter 'C' is **unlikely**.

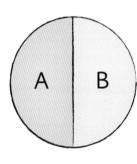

There is an **even chance** of spinning a letter 'A'.

There is an **even chance** of spinning a letter 'B'.

As area A is equal to area B, there is an equally likely chance of spinning 'A' as there is of spinning 'B'. This is the predicted probability. We may find that out of 10 spins we spin 'A' 7 times and 'B' 3 times. However, the more spins we make the more likely it is that we will spin an equal number of 'A's and 'B's.

Design a colour spinner that obeys all of these rules.

a Spinning blue is likely.

b Spinning green is unlikely.

c It is impossible to spin yellow.

d There is as much chance of spinning red as there is of spinning purple.

Now spin your spinner 50 times and record the colour spun each time. Does the frequency of each colour spun agree with the probabilities predicted above (**a**, **b**, **c** and **d**)? How do you know? If they are different, how would you explain this?

Statistics and Probability

Guided practice

Label the spinner so that:

• it is likely a 'W' will be spun

• it is unlikely an 'X' will be spun

• there is an even chance of spinning 'Y' or 'Z'.

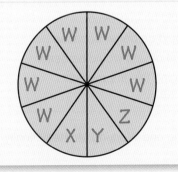

113

The Thinking and Working Mathematically Star

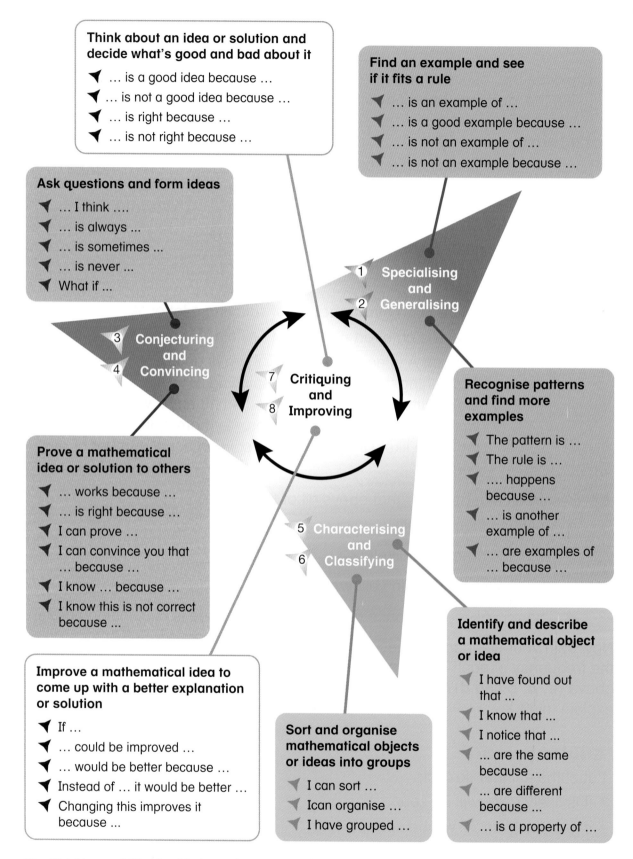

Think about an idea or solution and decide what's good and bad about it
- … is a good idea because …
- … is not a good idea because …
- … is right because …
- … is not right because …

Find an example and see if it fits a rule
- … is an example of …
- … is a good example because …
- … is not an example of …
- … is not an example because …

Ask questions and form ideas
- … I think ….
- … is always …
- … is sometimes …
- … is never …
- What if …

1 **2** Specialising and Generalising

3 **4** Conjecturing and Convincing

7 **8** Critiquing and Improving

Recognise patterns and find more examples
- The pattern is …
- The rule is …
- …. happens because …
- … is another example of …
- … are examples of … because …

Prove a mathematical idea or solution to others
- … works because …
- … is right because …
- I can prove …
- I can convince you that … because …
- I know … because …
- I know this is not correct because …

5 **6** Characterising and Classifying

Identify and describe a mathematical object or idea
- I have found out that …
- I know that …
- I notice that …
- … are the same because …
- … are different because …
- … is a property of …

Improve a mathematical idea to come up with a better explanation or solution
- If …
- … could be improved …
- … would be better because …
- Instead of … it would be better …
- Changing this improves it because …

Sort and organise mathematical objects or ideas into groups
- I can sort …
- Ican organise …
- I have grouped …

The Thinking and Working Mathematically star, © Cambridge International, 2018

Acknowledgements

Photo acknowledgements
Every effort has been made to trace copyright holders. Any omission will be rectified at the first opportunity.
p8 Pogorelova Olga/Shutterstock; p37t Stocksolutions/Shutterstock; p37b Avector/Shutterstock; p41 Maderla/Shutterstock; p42l MyPro/Shutterstock; p42r MichaelJayBerlin/Shutterstock; p51l MAHATHIR MOHD YASIN/Shutterstock; p51c Darrin Henry/Shutterstock; p51r Anna Nahabed/Shutterstock; p53 NikWB/Shutterstock; p59t MichaelJayBerlin/Shutterstock; p59b ZouZou/Shutterstock; p62 Jan Hyrman/Shutterstock; p78l Nadzin/Shutterstock; p78cl CGN089/Shutterstock; p78ccl Rawpixel.com/Shutterstock; p78ccr Germanskydiver/Shutterstock; p78cr Niwet Kumphet/Shutterstock; p78r Syda Productions/Shutterstock; p82 Toponium/Shutterstock; p85 Xpixel/Shutterstock; p91 Mallinka/Shutterstock; p114 Nattiyapp/Shutterstock.